SpringerBriefs in Philosophy

For further volumes:
http://www.springer.com/series/10082

Sabine Roeser · Rafaela Hillerbrand
Per Sandin · Martin Peterson
Editors

Essentials of Risk Theory

 Springer

Editors
Sabine Roeser
Department of Philosophy
Delft University of Technology
Delft
The Netherlands

Rafaela Hillerbrand
Department of Philosophy
Delft University of Technology
Delft
The Netherlands

Per Sandin
Department of Plant Physiology
   and Forest Genetics
Swedish University of Agricultural
   Sciences
Uppsala
Sweden

Martin Peterson
Section for Philosophy and Ethics
Eindhoven University of Technology
Eindhoven
The Netherlands

ISSN 2211-4548   ISSN 2211-4556 (electronic)
ISBN 978-94-007-5454-6   ISBN 978-94-007-5455-3   (eBook)
DOI 10.1007/978-94-007-5455-3
Springer Dordrecht Heidelberg New York London

Library of Congress Control Number: 2012947395

Printed on acid-free paper

Springer is part of Springer Science+Business Media (www.springer.com)

Preface

Risk has become one of the main topics in fields as diverse as engineering, medicine and economics, and it is also studied by social scientists, psychologists and legal scholars. But the topic of risk also leads to more fundamental questions such as: What is risk? What can decision theory contribute to the analysis of risk? What does the human perception of risk mean for society? How should we judge whether a risk is morally acceptable or not? Over the last couple of decades questions like these have attracted interest from philosophers and other scholars into risk theory.

This book presents the *Essentials of Risk Theory*. It is based on the more extensive *Handbook of Risk Theory*. The introductory chapter of the *Essentials of Risk Theory* book provides an overview of the full length *Handbook*. The other chapters are representative for the broad ranges of issues addressed in the *Handbook*. In this way, the *Essentials of Risk Theory* provides for a compact, easily accessible introduction in the various facets of the broad new field of Risk Theory. This makes the *Essentials of Risk Theory* an ideal text book for undergraduate and graduate courses on risk for fields as diverse as psychology, social sciences, economics, decision theory, and ethics.

The contributions are accessibly written and highly relevant to issues that are studied by risk scholars. We hope that the *Essentials of Risk Theory* and the *Handbook of Risk Theory* will be helpful starting points for students as much as for risk scholars who are interested in broadening and deepening their current perspectives.

Sabine Roeser (Editor-in-Chief)
Rafaela Hillerbrand
Per Sandin
Martin Peterson

Acknowledgments

Sabine Roeser's work for the *Handbook of Risk Theory* has been conducted at the philosophy departments of TU Delft and Twente University and was sponsored by The Netherlands Organization for Scientific Research (NWO), with VIDI-grant number 276-20-012.

Rafaela Hillerbrand's work on this volume was supported by the excellence initiative of the German federal and state governments and conducted at the Human Technology Centre (HumTec) and the Institute of Philosophy, RWTH Aachen University. Thanks to all members of the research group eet—ethics for energy technology for insights into risks from various disciplines. Special thanks to Andreas Pfennig, Nick Shackel, and Peter Taylor for numerous fruitful discussions on the topic.

Per Sandin's work for this handbook was done in the Department of Plant Physiology and Forest Genetics, Swedish University of Agricultural Sciences, Uppsala and in the Department of Philosophy and History of Technology, Royal Institute of Technology, Stockholm.

Martin Peterson has conducted his work for the handbook at the philosophy department of TU Eindhoven. Sabine Roeser and Martin Peterson are members of the 3TU. Centre for Ethics and Technology, a center of excellence of the federation of the three technical universities in The Netherlands (Delft, Eindhoven, and Twente).

We are very grateful to the contributors of this handbook. In addition, we would like to thank Katie Steele, Linda Soneryd, and Misse Wester, each of whom provided expert reviews for a chapter. We would like to thank the staff at Springer for the excellent collaboration, specially Ties Nijssen, Jutta Jaeger-Hamers, Christi Lue, and Christine Hausmann.

Contents

Contributors

Jessica Nihlén Fahlquist Delft University of Technology, Delft, The Netherlands; Royal Institute of Technology, Stockholm, Sweden

Melissa L. Finucane East-West Center, Honolulu, HI, USA

Rafaela Hillerbrand Department of Philosophy, Delft University of Technology, Delft, The Netherlands

Rolf Lidskog Centre for Urban and Regional Studies, Örebro University, Örebro, Sweden

Martin Peterson Section for Philosophy and Ethics, Eindhoven University of Technology, Eindhoven, The Netherlands

Hauke Riesch School of Social Sciences, Brunel University, London, UK

Sabine Roeser Department of Philosophy, Delft University of Technology, Delft, The Netherlands; Philosophy Department, University of Twente, Enschede, The Netherlands

Per Sandin Department of Plant Physiology and Forest Genetics, Swedish University of Agricultural Sciences, Uppsala, Sweden

Göran Sundqvist Centre for Technology, Innovation and Culture, University of Oslo, Oslo, Norway

Ibo van de Poel Delft University of Technology, Delft, The Netherlands

Chapter 1
Introduction to Risk Theory

Sabine Roeser, Rafaela Hillerbrand, Per Sandin and Martin Peterson

Introduction

Risk is an important topic in contemporary society. People are confronted with risks from financial markets, nuclear power plants, natural disasters and privacy leaks in ICT systems, to mention just a few of a sheer endless list of areas in which uncertainty and risk of harm play an important role. It is in that sense not surprising that risk is studied in fields as diverse as mathematics and natural sciences but also psychology, economics, sociology, cultural studies, and philosophy. The topic of risk gives rise to concrete problems that require empirical investigations, but these empirical investigations need to be structured by theoretical frameworks. This handbook offers an overview of different approaches to risk theory, ranging from general issues in risk theory to risk in practice, from mathematical approaches in decision theory to empirical research of risk perception, to theories of risk ethics and to frameworks on how to arrange society in order to deal appropriately with risk.

Risk theory provides frameworks that can contribute to mitigating risks, coming to grips with uncertainty, and offering ways to organize society in such a way that the unexpected and unknown can be anticipated or at least dealt with in a reasonable and ethically acceptable way. This Book presents essential topics in risk

S. Roeser (✉) · R. Hillerbrand
Department of Philosophy, Delft University of Technology, Delft, The Netherlands
e-mail: s.roeser@tudelft.nl

S. Roeser
University of Twente, Enschede, The Netherlands

P. Sandin
Swedish University of Agricultural Sciences, Uppsala, Sweden

M. Peterson
Eindhoven University of Technology, Eindhoven, The Netherlands

S. Roeser et al. (eds.), *Essentials of Risk Theory*, SpringerBriefs in Philosophy,
DOI: 10.1007/978-94-007-5455-3_1, © The Author(s) 2013

theory. It is based on a longer book, the Handbook of Risk Theory (Springer, 2012). The Handbook of Risk Theory reflects the current state of the art in risk theory, by bringing together scholars from various disciplines who review the topic of risk from different angles. In the following sections we provide for a short overview of the contributions to the Handbook of Risk Theory.

Part 1: General Issues in Risk Theory

Theoretical reflection about risk gives rise to various general issues. What is the relation between risk research and philosophy? What can the two disciplines learn from each other? How are the concepts of risk and safety interrelated? Which different levels of uncertainty can be distinguished? Part one of the handbook discusses these general issues in risk theory.

Sven Ove Hansson: A Panorama of the Philosophy of Risk

Sven Ove Hansson's chapter provides for an overview of the contributions that different philosophical subdisciplines and risk theory can provide for each other. He starts out with discussing the potential contributions of philosophy to risk theory. These contributions concern terminological clarification, argumentation theory, and the fact–value distinction. It is philosophy's "core business" to provide for terminological clarification. In the area of risk theory, philosophical theories can shed light on the multifaceted concepts of risk and safety. Philosophical argumentation theory can draw attention to common fallacies in reasoning about risk. In the philosophical tradition, there are intricate debates about the relationship between facts and values that can contribute to more careful and nuanced discussions about facts and values in risk analysis, for example, by making implicit value judgments explicit. These are areas in which existing philosophical theories can be applied rather straightforwardly. However, there are other issues in risk theory that give rise to new philosophical problems and require new philosophical approaches in virtually all areas of philosophy. This is due to the fact that most traditional philosophical approaches are based on deterministic assumptions. Thinking about risk and uncertainty requires radically different philosophical theories. Hence, the topic of risk can lead to new philosophical theories that are at the same time directly practically relevant, as our real-life world is one that is characterized by risk and uncertainty.

Niklas Möller: The Concepts of Risk and Safety

Niklas Möller provides for an analysis of the concepts of risk and safety. Möller distinguishes three major approaches in empirically oriented risk theory: the scientific, the psychological, and the cultural approach, respectively. Philosophical approaches connect with each of these approaches in specific ways. Möller then distinguishes between at least five common usages of the notion ''risk'' and shows how each usage can be connected to a different approach in risk research. He emphasizes the importance of distinguishing between factual and normative uses of the notion of risk. Möller continues with the question whether safety is the antonym of risk. He argues that this is not necessarily the case; for example, these words can have different connotations. Möller then discusses various ethical aspects of risk that are elaborated on in more detail in Part 5 of the handbook, on Risk Ethics. Möller's unique contribution to risk ethics is to argue that risk is a ''thick concept''; that is, a concept that does not only have descriptive aspects that are the subject of scientific investigations, but that also has normative or evaluative aspects, which require ethical reflection. Möller discusses and rejects, on philosophical grounds, various claims by social scientists to the socially constructed nature of risk that is supposed to follow from its inherently normative nature.

Hauke Riesch: Levels of Uncertainty

Risk science seems to be a paradigm of interdisciplinary research: Risk unites disciplines. But every discipline seems to denote something else with the umbrella term risk. This dilemma is the starting point of Hauke Riesch's contribution on the "Levels of Uncertainty". Hauke Riesch analyzes various uses of the terms risk and uncertainty. He attributes differences not so much to imperfect or sloppy use of the terms. Rather, he argues that these differences are a symptom of the fact that different scientists are interested in different aspects of risk. Therefore, there is not much point in criticizing someone for using vague or different notions of risk. Riesch conceptualizes risk as uncertainty of an event happening whose outcome may be severe. Riesch argues that concerning the uncertainty aspect of risk we can distinguish the following six questions which are not mutually independent: Why are we uncertain? Who is uncertain? How is uncertainty represented? How do people react to uncertainty? How do we understand uncertainty? What exactly are we uncertain about? Within this multidimensional map, Riesch divides the objects of uncertainty into five layers: uncertainty of the outcome, uncertainty about the parameters as well as uncertainty about the model itself, uncertainty about acknowledged inadequacies and implicitly made assumptions, and uncertainty about the unknown inadequacies. These layers relate to different concerns of different disciplines. The expert discourse commonly focuses on one of these levels, but this distorts the way people perceive particular risks, because higher level uncertainties still exist. Riesch illustrates through various case studies—lottery, bad eating habits, CCS (carbon capture storage), and climate change—how all five levels of uncertainty are always present, but differently important. Riesch's multidimensional classification provides some useful information for risk communication in order to convey information on other levels of uncertainty that people find important.

Part 2: Specific Risks

What brings about reflections on risk is the necessity to react to real natural or anthropogenic hazards. The second part of the handbook addresses natural, technological, and societal risks from the perspective of the natural and social sciences by also incorporating insights from the humanities and mathematical sciences.

Louis Eeckhoudt and Henri Loubergé: The Economics of Risk: A (Partial) Survey

Louis Eeckhoudt and Henri Loubergé give a historical overview of how risk thinking has developed in the mainstream model of economics, that is, that of expected utility theory. The authors argue that despite Daniel Bernoulli's foundational work on this model in the eighteenth century, those ideas were not formulated in precise terms until von Neumann and Morgenstern's Expected Utility Theorem in the late 1940 s. These ideas were further developed by Friedman and Savage, Arrow, and Pratt. Eeckhoudt and Loubergé introduce the concept of general equilibrium and discuss issues of risk distribution among individuals. The authors end their discussion with brief illustrations of applications of the theory from the fields of finance and insurance.

Reinoud D. Stoel and Marjan Sjerps: Interpretation of Forensic Evidence

Reinoud D. Stoel and Marjan Sjerps write about the interpretation of forensic evidence. Since absolute certainty regarding the guilt of a suspect is unattainable, the question has to be put in terms of probability. They go on to describe how this issue can be approached using the Likelihood Ratio and how this applies to both forensic experts and legal decision makers. They propose further research focusing on methods for computing quantitative Likelihood Ratios for different forensic disciplines and also how different pieces of evidence combine, for example, using Bayesian networks.

John Weckert: Risks and Scientific Responsibilities in Nanotechnology

John Weckert writes about the risks of nanotechnology, and using this field as an example he discusses scientists' responsibility. He presents four generations of nanotechnology and the risks associated with each generation. In particular, he emphasizes five risks that have been discussed in relation to nanotechnology: health/environmental risks related to nanoparticles, "grey goo," threats to privacy, cyborgs, and the possibility of a "nanodivide" between the developed and the developing world. He goes on to delineate two models of science, the linear and the social mode, where the latter focuses on the inevitable value-ladenness of science. He also discusses four different interfaces between science and ethics. As regards the responsibility of scientists involved in nano research, he argues that these responsibilities differ between the different risks, and that on the linear model scientists are not free from responsibility.Weckert calls for more interdisciplinary research in the future, based on sound science and knowledge of actual techno-logical developments.

Annette Rid: Risk and Risk-Benefit Evaluations in Biomedical Research

Annette Rid critically reviews recent debates about risk-benefit evaluations in biomedical research. To determine whether potential new interventions in clinical care, new drugs, or basic science research in biomedicine have a net benefit, risks that participants take have to be balanced against the positive effects for potential patients and society as a whole. Rid presents and evaluates the four existing ethical frameworks for risk-benefit evaluations in medical ethics: component analysis, the integrative approach, the agreement principle, and the net risks test. She contends that net risks tests are superior to alternative approaches, but fail to offer guidance for evaluating the ethical acceptability of risks that participants are exposed to for research purposes only. This leaves two of the fundamental problems of risk-benefit evaluations in research largely unaddressed: How to weigh the risks to the individual research participant against the potential social value of the knowledge to be gained from a study, and how to set upper limits of acceptable research risk. Discussions about the "minimal" risk threshold in research with participants who cannot consent go some way to specifying upper risk limits in this context. However, these discussions apply only to a small portion of research studies.

Erika Mansnerus: Understanding and Governing Public Health Risks by Modeling

Erika Mansnerus discusses how risk is perceived and expressed though the computational models that are increasingly used to govern and understand public health risk. She discusses a case study on infectious disease epidemiology to illustrate how models are used for explanation- based and scenario-building predictions in order to anticipate the risks of infections. Mansnerus analyzes the tension that arises when model-based estimates exemplify the population- level reasoning of public health risks, but have restricted capacity to address risks on an individual level. Mansnerus provides an alternative account in order to overcome the limitations of computational tools in the governance of public health risks. This richer picture on risk goes beyond the pure probabilistic realm of mathematical risk modeling.

John Adams: Management of the Risks of Transport

John Adams discusses whether transport safety regulations indeed reduce risks of, for example, fatal car accidents. With the exception of the seat belt law, major reductions in road fatalities over the past decades cannot be linked to changes in transport safety laws in a straightforward way. So what is the use of these laws and why do they fail? Adams introduces the reader to two models of how risk managers may deal with risks, a cost–benefit model and a model underlying risk compensation strategies. Both models focus on different risk. The former focuses on what Adams terms "risks perceived through science" and models road users as ignorant, obedient objects. In contrast, the alternative model pictures road users as vigilant and responsive subjects and thus can also take into account directly perceived risks and what Adams terms "virtual risks." Adams applies Mary Douglas' cultural theory of risk to road safety issues in order to provide an explanation of why different societies perceive risks very differently. Adams shows how this leads to two very different strategies to manage and reduce the risks of transport in the two countries with the best road safety records worldwide, the Netherlands and Sweden.

Claudia Basta: Risk and Spatial Planning

Claudia Basta challenges the way societies currently deal with site-specific hazardous technologies. The location of hazardous facilities is rarely an uncontroversial issue. Basta describes how this issue is dealt with in selected European countries and explains their differences as a reflection of nonexplicit, but retraceable, underlying ethical theories. Basta suggests a possible synergy between spatial planning practices and ethical theories by proposing a theoretical framework which may guide spatial planning processes before and beyond unavoidable contextual features. In particular, following the Rawlsian theory of justice (Rawls 1971) as transposed to spatial planning theories by Moroni, Basta proposes an understanding of "spatial safety" as a primary spatial good. Spatial planning practice is thus perceived primarily as a practice of distributing the maximum possible amount of the primary good of spatial safety in society equally up to the lowest societal level. She contends that approaching any siting controversy as a NIMBY (not-in-my-backyard) case is not only incorrect, but also dangerously instrumental. The case study of a planned CCS facility at Barendrecht illustrates this point.

Behnam Taebi: Intergenerational Risks of Nuclear Energy

Behnam Taebi discusses the intergenerational risk of nuclear waste disposal. The accident in the Fukushima II nuclear power plant in Japan in 2011 gives rise to a renewed discussion about the civic use of nuclear power. In addition to the risk of a nuclear meltdown, nuclear power plants put seemingly unduly high burdens on future generations: The longevity of the toxic and radioactive waste seems to require geological disposals that are faced with immeasurable uncertainties concerning the stability of rock formation over large timescales (a few thousand years). The relatively new technological possibility of Partition and Transmutation (PT) may provide an alternative to that and put long-term surface storage in a new light. Taebi not only examines this new technology as an alternative to geological disposal, but also addresses central aspects of intergenerational ethics, namely the principles of intergenerational equity. Taebi scrutinizes the notion of diminishing responsibility over time, which is an important notion in nuclear waste policies, and rejects the moral legitimacy of distinguishing between future people. In this sense Taebi not only provides a thorough account on how to deal with longlife radioactive waste, but also reflects on our present obligations toward future generations.

Rafaela Hillerbrand: Climate Change as Risk?

Rafaela Hillerbrand analyzes the types of uncertainties involved in climate modeling and discusses whether common decision approaches based on the precautionary principle or on the maximization of expected utility are capable of incorporating the uncertainty inevitably involved in climate modeling. The author contends that in the case of decision making about climate change, unquantified uncertainties can neither be ignored, nor can they be reduced to quantified uncertainties, by assigning subjective probabilities. These insights reveal central problems, as they imply that the commonly used elementary as well as probabilistic decision approaches are not applicable in this case. The chapter argues that the epistemic problems involved in modeling the climate system are generic for modeling complex systems. Possible parts of future research to circumvent these problems are adumbrated.

Amy Donovan: Earthquakes and Volcanoes: Risk from Geophysical Hazards

The recent earthquake in Japan revealed the vulnerability of a high-tech society to natural hazards. Amy Donovan calls for a genuinely interdisciplinary study of volcano and seismic risk if we want to be better forewarned against such risks. A wide range of disciplines, spanning both social and physical sciences, is involved in research into geophysical hazards, in order to predict, prepare for, and communicate about events like the one in Japan. However, only few holistic approaches to geophysical risks exist. Most information comes from the natural sciences that tend to focus on hazard assessment, while the social sciences focus on vulnerability reduction and risk communication. The contribution of Amy Donovan fills this gap. Donovan examines the social context both of the scientific research of the natural hazards, and of the hazards themselves, proposing a holistic and context-based approach to understanding risk. In the context of seismic risk, the uncertainty of the scientific methods coupled with the procedures involved in mitigating these risks and the need to involve populations in the preparation, leads to a snowballing of uncertainty and indeterminacy from the scientific domain through the policy domain and into the wider public.

Part 3: Decision Theory and Risk

As illustrated by the interdisciplinary nature of the contributions to the handbook, risk theory is a very broad field of research. Eight chapters explore the links between risk theory and *decision theory*.

Very briefly put, decision theory is the theory of rational decision making. A key assumption accepted by nearly all decision theorists is that it is essential to distinguish between *descriptive* claims about how people actually make risky decisions, and *normative* claims about how it would be rational to take such decisions. In recent years, the term decision theory has primarily been used for referring to the study of normative claims about rational decision making. In what follows, we shall follow this convention and reserve the term "decision theory" for the study of normative hypotheses about what rationality demands of us.

The notion of rationality researched in decision theory is best described as means—ends rationality. The rational decision maker has certain beliefs about what the effects of various risky options could be, and also has a set of desires about what he or she wants to achieve. The key question is then how we should best combine these beliefs and desires into a decision.

This so-called belief-desire model of rationality goes back at least to the Scottish eighteenth- century philosopher David Hume. According to Hume, the best explanation of why people behave as they do—and are willing to accept the risks they do accept—is that we all have certain beliefs and desires that determine our actions (cf. Hume 1740/1967). In the belief-desire model, a basic criterion of a rational decision is that it must accurately reflect the agent's beliefs and desires, no matter what these beliefs and desires happen to be about. But how should such a theory of rationally permissible structures of beliefs and desires be spelled out in detail?

The mainstream view among contemporary decision theorists is that rational agents are allowed to let whatever beliefs and desires they so wish guide their decisions, as long as those beliefs and desires are compatible with the principle of maximizing expected utility. The principle of maximizing expected utility takes the total value of an act to equal the sum total of the values of its possible outcomes weighted by the probability for each outcome. The values assigned to an outcome are determined by the decision maker's desires, whereas the probabilities are determined by his or her beliefs about how likely the outcomes are to materialize.

An important step in the development of modern decision theory was the development of axiomatic accounts of the principle of maximizing expected utility. The first such axiomatization was sketched by Frank Ramsey in his paper *Truth and Probability*, written in 1926 but published posthumously in 1931. Ramsey (1931) formulated eight axioms for how rational decision makers should form preferences among uncertain prospects. One of the many points in his paper is that a decision maker who behaves in accordance with the eight axioms will act in a way that is *compatible* with the principle of maximizing expected value, in the sense that he or she will implicitly assign numerical probabilities and values to outcomes. Note that it does not follow from this that the decision maker's choices were *actually triggered* by these implicit probabilities and utilities. This indirect approach to rational decision making is extremely influential in the contemporary literature. In 1954, Leonard Savage put forward roughly the same idea in his influential book *The Foundations of Statistics* (Savage 1954).

An equally important book in the recent history of decision theory is John von Neumann and Oskar Morgenstern's book *Theory of Games and Economic Behavior* (1944). They used the notion of a "lottery" for developing a linear measure of an outcome's utility. In von Neumann and Morgenstern's vocabulary, a lottery is a probabilistic mixture of outcomes. For instance, an entity such as "a fifty-fifty chance of winning either $1,000 or a trip to Miami" is a lottery. The upshot of their utility theory is that every decision maker whose preferences over a very large set of lotteries conform to a small set of axioms implicitly assigns numerical utilities to outcomes, and also implicitly acts in accordance with the principle of maximizing expected utility.

The chapters in this section discuss various aspects of, ideas behind, and problems with decision theory in the context of risk.

Claus Beisbart: A Rational Approach to Risk?
Bayesian Decision Theory

Claus Beisbart's contribution clarifies the concept of utility maximization as the core notion of rationality in Bayesian decision theory. For Bayesians, a rational approach to risk depends on the agents' utilities and subjective probabilities that measure the strength of their desires and beliefs, respectively. Two questions are commonly put forward by critics of Bayesian decision theory: Why do or why should rational agents maximize expected utility? And how can the strength of an agent's desire and belief be measured? The classical answers by Bayesians to these questions are commonly put in the form of representation theorems. These theorems show that – under certain assumptions – an agent's beliefs and desires can indeed be represented in terms of numerical probabilities and utilities. Beisbart presents several different ways to obtain a presentation theorem: the classical approach of von Neumann-Morgenstern, its modification by Anscomb and Aumann, the approach of Ramsey, Bolker-Jeffrey theory, and Savage's account, the latter possibly being closest to how risk is actually dealt with in technology assessment. The chapter concludes with a discussion of major controversies concerning Bayesian decision theory.

Karsten Klint Jensen: A Philosophical Assessment
of Decision Theory

Karsten Klint Jensen starts his chapter with drawing a distinction between classical decision theory and modern axiomatic decision theory. He then goes on to give an overview of Savage's axiomatization of the principle of maximizing expected utility. This all leads up to a discussion of what *sort of problem* decision theory really aims at solving. Several possible answers are considered. Jensen points out that the perhaps most plausible interpretation of what modern decision theories are up to is to try to develop theories of what counts as personal good and how such personal goods can be aggregated intrapersonally.

Peter R. Taylor: The Mismeasure of Risk

Peter R. Taylor challenges the classical approaches in risk management and risk assessment, thereby mainly, but not exclusively, focusing on how the insurance industry deals with risk. Taylor argues that already our embosomed definition of risk, which pictures risk as a quantitative measure combining, in one way or the other, harm and likelihood of the hazardous event, falls short of describing realistic events like the recent disasters that caught world headlines – tsunamis, volcanic ash clouds, or financial crashes. It is argued that simple measures of risk that, like for example in the Bayesian approach, focus merely on the mean expected harm, may be poor guides for dealing with real-world risk. Taylor outlines how a more complex risk assessment may incorporate what Rumsfeld termed the unknown unknowns or Taleb the black swans: that is, events that are not considered in the probability space. Taylor shows how model risk, that is, the risk that the underlying model of the, say, physical hazard is simply inadequate, may be quantified and how multiple measures and thresholds may be implemented. The author further examines if we could also tackle what he calls ROB, "risk outside the box," that is, risks that are not considered by the underlying risk model.

Nils-Eric Sahlin: Unreliable Probabilities, Paradoxes, and Epistemic Risks

Nils-Eric Sahlin's chapter focuses on the quality of the information on which a decision is based. By definition, if the information at hand when a decision is taken is unreliable, then the epistemic uncertainty is high. Suppose, for instance, that for one reason or another tomorrow's weather is important for a decision that you are about to take. Now consider the scenario in which you have received a detailed weather forecast from a professional meteorologist according to which the probability for rain tomorrow is 50% and compare this with a scenario in which your 3-year-old son, who knows nothing about meteorology, tells you that he thinks the probability for rain tomorrow is 50%. Intuitively, you as a decision maker seem to be better off from an epistemic point of view in the first scenario compared to the second, but exactly how should this difference be accounted for from a decision theoretical point of view? How do we take intuitions about epistemic risks into account in our theories of rational decision making? This is the key question that drives Sahlin's chapter.

Till Grüne-Yanoff: Paradoxes of Rational Choice Theory

Till Grüne-Yanoff 's chapter gives an overview of some of the many well-known paradoxes that have played an important role in the development of decision theory. The very first such paradox was the St Petersburg paradox, formulated by the Swiss mathematician Daniel Bernoulli (1700–1782), who worked in St Petersburg for a couple of years at the beginning of the eighteenth century. The St Petersburg paradox, which is still being discussed by decision theorists, is derived from a game known as the St Petersburg game: A fair coin is tossed until it lands heads up. The player then wins 2^n dollars, where n is the number of times the coin was tossed. Hence, if the coin lands heads up in the first toss, the player wins 2 dollars, but if it lands heads up on, say, the fourth toss, the player wins $2^4 = 2 \cdot 2 \cdot 2 \cdot 2 = 16$ dollars. How much should a rational person be willing to pay for getting the opportunity to play this game? Clearly, the expected monetary payoff is infinite, because $\frac{1}{2} \cdot 2 + \frac{1}{4} \cdot 4 + \frac{1}{8} \cdot 8 + \ldots = 1 + 1 + 1 + \ldots = \sum_{n=1}^{\infty} \left(\frac{1}{2}\right)^n \cdot 2^n = \infty$. However, to pay, say, a million dollars for playing a game in which one is very likely to win just 2, 4, or 8 dollars seems absurd. As Grüne-Yanoff points out, it is not as easy to resolve this and other paradoxes as many people have thought.

Paul Weirich: Multi-Attribute Approaches to Risk

Paul Weirich considers the special type of situation that arises if many different features of a decision are considered to be relevant by the decision maker. In a single-attribute approach to decision theory, all possible outcomes are compared on one and the same scale. Imagine, for instance, that you are about to buy a new car. Some cars are more expensive than others, but they are also safer. How many dollars is it worth to pay for extra safety? The multi-attribute approach attempts to avoid the idea that money and human welfare are somehow directly commensurable by giving up the assumption that all outcomes have to be compared on a common scale. In a multi-attribute approach, each type of attribute is measured in the unit the decision maker considers to be most suitable for the attribute in question. Money is typically the right unit to use for measuring financial costs, whereas other measures are required for measuring car safety.

John R. Welch: Real-Life Decisions and Decision Theory

John R.Welch discusses another important area of decision theory, concerning the question of how we can apply decision theory to real-life decisions. As everyone who reads the other chapters on decision theory will quickly discover, decision theorists make many quite unrealistic idealizations about the decision problems they are discussing, which seldom or never hold true in real-life applications. In recent years, decision theorists have become increasingly aware of this limitation, and in response to this have started to develop more realistic decision theories that deal better with the ways in which we actually take decisions in real life. Welch gives a very instructive overview of this emerging literature.

Stefan T. Trautmann and Ferdinand M. Vieider: Social Influences on Risk Attitudes: Applications in Economics

In the last chapter on decision theory, Stefan T. Trautmann and Ferdinand M. Vieider discuss the links between decision theory and two of the key disciplines on which much work in decision theory draws, namely, economic theory and psychology. Trautmann and Vieider identify and discuss four distinct types of social influences on economic decisions under risk: (1) observations of other agents' outcomes; (2) observations of the decision maker's outcomes by other agents; (3) direct effects of the decision maker's choices on other agents' outcomes; and (4) direct dependencies of the decision maker's outcomes on other agents' choices.

Part 4: Risk Perception

Where the previous section discussed normative or rational decision theory, the present section discusses risk from the point of view of empirical decision theory, or to use a more common notion, risk perception. In the 1970s, the psychologists Amos Tversky, Daniel Kahneman, and Paul Slovic started to investigate the ways in which people as a matter of fact make decisions under uncertainty or risk. It turned out that these decisions deviate significantly from rational decision theory. It might not be too surprising that laypeople's intuitive judgments about risk and statistics deviate from mathematical methods. However, surprisingly, also the judgments of experts turned out to be subject to numerous heuristics and biases. These findings gave rise to a whole research industry in mistakes people make in their risk judgments (cf., e.g., Gilovich et al. 2002), which eventually earned Kahneman the Nobel Prize in economics in 2002. The common framework to explain these phenomena is Dual Process Theory (DPT) which states that there are two systems with which people make judgments: system 1 is intuitive, spontaneous, and evolutionary prior; system 2 is rational, analytical, and comes later in our evolution. System 1 helps us to navigate smoothly through a complex world, but system 2 is the one that provides us with ultimate normative justification.

However, there are alternative approaches to risk perception that challenge this picture to some extent. Paul Slovic and various social scientists have argued that there is no "objective" measure of risk, that all approaches to risk, also those of experts, involve normative and partially arbitrary or subjective assumptions. Slovic and his colleagues have conducted studies that show that laypeople do not so much have a wrong understanding of risk but rather a different understanding of risk that might provide for valuable insights (cf. Slovic 2000). Dan Kahan has combined Slovic's psychometric approach with Mary Douglas's cultural theory to account for cultural values in risk perception. The psychologist Gerd Gigerenzer (2007) has conducted studies that show that intuitive risk judgments can actually be more reliable than mathematical approaches to risk. For example, experts' intuitions, but also laypeople's heuristics, can be superior to formal approaches.

This section consists of chapters that discuss various aspects of risk perception.

Dylan Evans: Risk Intelligence

Dylan Evans presents new empirical research that shows that risk intelligence perceived as the ability to estimate probabilities correctly is rare. Previous calibration tests have mainly been used to measure expert groups like medics and weather forecasters. However, Dylan presents tests of over 6,000 people of all ages and a variety of backgrounds and countries. Like other work in the psychology of judgment and decision making, Evans' own work shows that most people are not very good at thinking clearly about risky choices. They often disregard probability entirely, and even when they do take probability into account, they make many errors when estimating it. However, some groups of people have an unusually high level of risk intelligence. Evans outlines how lessons can be drawn from these groups to develop new tools to enhance risk intelligence in others.

Nicolai Bodemer and Wolfgang Gaissmaier: Risk Communication in Health

The chapter by Nicolai Bodemer and Wolfgang Gaissmaier studies risk communication in the health-care sector. A major problem for doctors who wish to help their patients to make wellinformed medical decisions is that patients often find it difficult to understand the information presented by the doctor. Bodemer and Gaissmaier point out that purely qualitative information, such as saying that the risk is "large" or "quite small," often does not work well: the beliefs that the patients end up with if they receive purely qualitative information is often badly calibrated with the true risk. Numerical representations often make it easier for patients to correctly understand the magnitude of a risk. However, some numerical representations tend to be easier to understand that others. In particular, Bodemer and Gaissmaier argue that doctors should try to avoid using conditional probabilities if they wish to be understood, and instead use natural frequencies.

Lennart Sjöberg: Risk Perception and Societal Response

Lennart Sjöberg discusses research on risk perceptions of experts and laypeople. Sjöberg first reviews the most well-known models of risk perception, that is, the psychometric model by Paul Slovic and others, and Cultural Theory as developed by Mary Douglas and Aaron Wildavsky (1982). Sjöberg argues that the empirical evidence for these models is problematic. These approaches leave 80% of the variance in risk perception unexplained. Sjöberg presents approaches that have more explanatory power. One tool is the risk sensitivity index with which risk attitudes can be measured. Sjöberg goes on to discuss the role of affect and emotion in risk perception. He points out ambiguities in the use of these notions in studies on risk perception; for example, affect sometimes refers to emotions and sometimes to values. He emphasizes that the psychometric model only employs one emotion, that is, dread, and it is rated for others rather than for the respondent. Sjöberg also discusses social trust, epistemic trust, and antagonism as important dimensions in risk perceptions. He reviews studies that show that risk perceptions of experts and laypeople mainly diverge when experts have responsibilities for risk, which can be explained by self-selection and social validation that lead to lower risk perceptions amongst experts. Sjöberg emphasizes that there will probably never be an ultimate consensus on risk in an open society. He warns for the risks of risk denial.

Melissa L. Finucane: The Role of Feelings in Perceived Risk

Melissa L. Finucane discusses the role feelings play in risk perception. She starts with a historical review of the development in research of feeling in risk perception, which is usually placed within the framework of Dual Process Theory. She then discusses different functions of feelings that have been identified in the context of risk, such as providing for accessible information, motivation, and moral and evaluative knowledge. Finucane presents frameworks that question the dichotomies underlying Dual Process Theory and provides for alternative views, for example, that "risk as analysis" (system 2) and "risk as feeling" (system 1) can be combined into what Finucane calls "risk as value". Finucane goes on by reviewing empirical evidence for the role of feelings in risk perception. One example is the role of emotional images on risk perception. Another example is psychophysical numbing, referring to diminishing sensitivity as numbers of, for example, victims of a disaster increase. This phenomenon can explain why we fail to respond appropriately to large humanitarian or environmental disasters. Finucane also discusses biases in gambles that are due to feelings, and the influence of moods on risk perception. She concludes by pointing out directions for future research, by taking into account available empirical tools for research into feelings, and alternative approaches to risk that go beyond purely quantitative models but also include values and feelings.

Ross Buck and Rebecca Ferrer: Emotion, Warnings, and the Ethics of Risk Communication

Ross Buck and Rebecca Ferrer discuss the relation between emotion, warnings, and the ethics of risk communication. They describe the common approach to risk communication which focuses on factual, statistical information. Appeal to emotions is considered to be unethical because it is supposed to be a form of manipulation. Buck and Ferrer challenge this approach. Emotions are already widely used in marketing, often overruling the more sober information on risk, for example, in the context of tobacco and alcohol consumption. They review empirical studies from decision theory and neuropsychology that show the importance of emotions in decision making. Some of these studies support the framework of Dual Process Theory, others suggest the possibility of an interaction between the affective and the analytical system, and yet others indicate that affect and cognition are intertwined. The authors argue that effective and ethically sound risk communication has to take into account and anticipate the various ways in which emotions can play a role in risk decisions. They present work on the role of emotion in communication about safe sexual behavior and on emotion intervention strategies and emotional education to illustrate this.

Dan M. Kahan: Cultural Cognition as a Conception of the Cultural Theory of Risk

Dan M. Kahan discusses two approaches to risk perception: Mary Douglas's and Aaron Wildavsky's cultural theory, and cultural cognition of risk. The latter is a combination of the former with Paul Slovic's psychometric approach to risk. Kahan first provides for a rough outline of cultural theory and its developments. According to cultural theory, cultural worldviews can be fit in a matrix with two axes called "group" and "grid." The group axis ranges from individualism to solidarity, the grid axis from hierarchy to egalitarianism. What distinguishes cultural cognition from other versions of cultural theory is that it allows for a certain way to measure cultural worldviews, a focus on the social and psychological measures that explain the way culture shapes risk perceptions, and a focus on practical applications. The chapter addresses each of these points. It reviews various studies by Kahan and colleagues that show how influential people's cultural worldviews are on the kinds of risks they find salient and which experts they find trustworthy. One way to mitigate that effect is to have experts proclaim unexpected viewpoints, for example, a leftist looking expert making typical right-wing claims. However, Kahan notes that to do this would be ethically dubious. A solution would be to include a plurality of viewpoints in public debates, as this also leads to less predictable views amongst people.

Britt-Marie Drottz-Sjöberg: Tools for Risk Communication

Britt-Marie Drottz-Sjöberg studies risk communication projects. She presents different risk communication tasks and derives several general conclusions.

Tools for risk communication combine theoretical and applied insights. Drottz-Sjöberg discusses examples which show the influences of social and historic events on a communication setting or a conflict situation and how they shape a risk communication project. She also analyzes how values, attitudes, and feelings influence thinking and behavior within groups. The examples provide for heuristics for the improvement of risk communication. Drottz-Sjöberg also discusses the RISCOM model of transparence. She shows that risk communication always takes place in a social setting, involving various interests, power relations, and actors' own agendas. However, the aim to communicate about specific risks, nevertheless, can be focused on clarification, understanding, and learning. Drottz-Sjöberg describes tools for risk communication that aim at achieving clarity in dialogues, which are characterized by openness and interaction regarding risk issues, in order to enhance problem solving and democracy.

Part 5: Risk Ethics

There is a growing consensus amongst risk scholars that risk is not a purely quantitative notion but also involves qualitative, normative, and ethical considerations. The dominant approach in risk analysis and risk management is to define risk as the probability of unwanted outcomes, such as annual fatalities, and to apply cost–benefit (or risk–benefit) analysis to determine which of various alternative technologies or activities is preferable. However, the question as to which unwanted outcomes to take into account already involves ethical considerations. Furthermore, cost–benefit analysis compares aggregates, whereas it is ethically significant how costs and benefits are distributed within a society. Social scientists and philosophers argue that ethical considerations such as justice, fairness, equity and autonomy have to be taken into account in assessing the acceptability of risk.

Interestingly, the same considerations can be found in the risk perceptions of laypeople that have been studied by Paul Slovic and others and which are discussed in the section on risk perception. Apparently, the intuitive responses to risk by laypeople include ethical aspects. The question arises why these considerations do not figure in the approaches of experts. This might be due to the fact that expert approaches are by definition focused on quantitative data and mathematical tools. Although these approaches can be helpful to a certain degree, they can lead to a tunnel vision that excludes other important considerations. In the context of risk it turns out that laypeople intuitively have a broader perspective that does justice to ethical considerations that can be normatively justified through established ethical theories.

The chapters in this section discuss ethical aspects of risk in more detail.

Douglas MacLean: Ethics and Risk

The point of departure in Douglas McLean's contribution is the widespread belief that traditional ethical theories have little, if anything, to say about risk. Numerous contemporary scholars argue that moral philosophers of the past have simply failed to recognize the ethical issues related to risk, and that this is therefore an area in which more theoretical work is needed. McLean claims that this mainstream picture is not entirely true: Although risk has not been one of the major topics of ethical reflection in the past, it is easy to find examples of scholars who have explicitly discussed ethical principles for risk decisions. The most prominent examples are John Stuart Mill, Jeremy Bentham, and Robert Nozick; the latter, for instance, devoted an entire chapter of his influential book *Anarchy, State, and Utopia* to ethical principles for risk decisions. This means that the ethics of risk has been extensively analyzed within at least two ethical traditions, namely, utilitarianism and theories based on natural rights.

Carl F. Cranor: Toward a Premarket Approach to Risk Assessment to Protect Children

Carl F. Cranor critically discusses risk legislation, specifically the postmarket approach to risk that is currently common in the USA and many other countries. In contrast with this, Cranor argues in favor of a premarket approach on which risks are assessed before products enter the market, similar to legislation concerning pharmaceuticals. Most industrial chemicals are allowed to enter the market without any testing. Tests are done afterwards, which often leads to strategic behavior, such as claims concerning supposed insufficient evidence about risks. This happened in the case of the tobacco industry, which delayed legislation against smoking for decades. For 70% of the chemicals used for products, there are no toxicity data at all. Cranor reviews evidence of significant amount of traces of dangerous industrial chemicals that can be found in the population. He specifically focuses on health risks for children to give special force to his argument. Given the special vulnerability of small children and fetuses, they should be given additional protection. Legislation should require testing before chemicals are used in consumer products. This will lead to a paradigm shift in legislation as much as in scientific practice.

Sabine Roeser: Moral Emotions as Guide to Acceptable Risk

Sabine Roeser explores the role emotions do and can play in debates about risky technologies. Most authors who write on risk and emotion see emotions as a threat to rational decision making about risks. These authors endorse Dual Process Theory, according to which emotion and reason are distinct faculties that have opposite tasks. However, based on recent developments in emotion research, an alternative picture of risk emotions is possible. According to various psychologists and philosophers who study emotions, emotions are a source of practical rationality. They are appraisals or judgments of value that have a cognitive aspect. These ideas can be applied to risk emotions. Emotions such as sympathy and compassion help to grasp morally salient aspects of risk, such as fairness, justice, and autonomy. This view allows for fruitful insights on how to improve public debates about risk, by taking emotional concerns of the public, but also of policy makers and experts, seriously. This approach leads to morally better judgments about risks, by doing justice to emotional-ethical concerns. In addition, as all parties will be taken seriously, it can also help to overcome the gap between experts and laypeople that currently so often leads to a deadlock in discussions about risky technologies.

Allison Ross and Nafsika Athanassoulis: Risk and Virtue Ethics

Allison Ross and Nafsika Athanassoulis propose a virtue ethics approach to risk assessment. They argue that it is superior to consequentialist or deontological approaches to the moral assessment of risk. For example, consequentialist approaches to risk are not sensitive to morally salient aspects of risk such as recklessness, fairness, and equity. Risk assessment cannot be left to scientific experts. Risk-taking is both unavoidable and potentially morally problematic. Hence, it requires context-sensitive and reflective judgments. Ross and Athanassoulis argue that it is important to focus on the role of character and patterns of behavior in moral risk assessments. Such patterns should not be understood as the result of arbitrary, automatic processes but as the product of dispositions which constitute somebody's character. Character dispositions are developed through education, habituation, and reflection. They combine desires, emotions, and thoughts that are attuned to decision making about risk in specific circumstances. The authors argue that only virtue ethics with its emphasis on character provides for a framework for sensible and reflective risk judgments. They illustrate this with a hypothetical time-travel experiment in which an agent has to decide about risks for himself or herself and others.

Philip J. Nickel and Krist Vaesen: Risk and Trust

Philip J. Nickel and Krist Vaesen discuss philosophical conceptions of the relationship between risk and trust. They distinguish between three main approaches.

The first is a Hobbesian approach. This approach understands trust as a kind of risk assessment about the expected behavior of other people and the estimated benefits of cooperation. This approach comes close to expected utility theory, which is commonly used in formal decision theoretical approaches to risk. The second approach to risk and trust is in direct opposition with such a calculative risk assessment. On this approach, one willingly relies on people based on, for example, habitual, social, or moral reasons. On the third approach, trust is seen as a morally loaded attitude, in which one expects the trusted person to fulfill certain moral obligations. This allows for cooperative behavior in which there are no interpersonal risks. Nickel and Vaesen examine how these three approaches explain relationships between the concepts of risk and trust, also based on empirical research, specifically on cooperative breeding. They suggest that the notion of trust might help overcome the current gap between technocratic and social approaches to risk, if experts are more aware of their moral responsibilities rather than simply providing the public with information, which might create fears.

Ibo van de Poel and Jessica Nihlén Fahlquist: Risk and Responsibility

Ibo van de Poel and Jessica Nihlén Fahlquist discuss the relationship between risk and responsibility. The authors start with noting that even though it is very common to link these two concepts in our daily practice, there is hardly any academic literature on it. They first discuss different conceptions of and connections between risk and responsibility. Van de Poel and Nihlén Fahlquist then elaborate on specific topics concerning responsibility for risk. They discuss the responsibility of engineers to contribute to risk reduction. They elaborate on the role of values and responsibility in risk assessment, risk management, and risk communication. An important distinction is the one between individual and collective responsibility for risk. The authors illustrate this with a case study from traffic safety. Van de Poel and Nihlén Fahlquist suggest various topics for future research, centered around the so-called problem of many hands in relation to climate change. The authors propose three possible lines of research to address this problem, which are responsibility as virtue, procedures for distributing responsibility, and institutional design.

Madeleine Hayenhjelm: What Is a Fair Distribution of Risk?

Madeleine Hayenhjelm's chapter discusses the question as to what a fair distribution of risk is, partially based on insights from John Rawls's theory of justice. Hayenhjelm starts out by reviewing what the objects of fairness are in the context of risk distributions. It is commonsensical that goods should be increased and risks should be diminished. This is also the underlying rationale of risk–cost–benefit analysis. However, such a consequentialist approach to risk overlooks issues of fair distribution, by only focusing on aggregate risks and benefits. Goods and risks should be distributed fairly. This can give rise to moral dilemmas. Hayenhjelm discusses problems with equal distributions of probabilities of harm by focusing on a thought experiment from James Lenman. She then reviews conditions under which deviations of equal distributions are justified and can still be fair. She suggests that this requires the justification of specific risky activities, and that higher risks for specific people should be mitigated by consent, precaution, and compensation.

Lauren Hartzell-Nichols: Intergenerational Risks

Lauren Hartzell-Nichols discusses the notion of intergenerational risks – long-term threats of harm that will affect future people – using the example of climate change. She begins by noting that there is comparatively little material on inter-generational risk, and identifies two philosophical problems that are relevant for the issue: Parfit's Non-Identity Problem and Gardiner's Pure Intergenerational Problem. She introduces a distinction between de re and de dicto badness that can be illuminating. Hartzell-Nichols presents three current approaches to addressing intergenerational risks: (1) cost–benefit analysis, (2) precautionary principles, and (3) approaches based on intergenerational justice, for example, as discussed by Darrel Moellendorf, Henry Shue, Simon Caney, and Steven Vanderheiden. She argues why the two former approaches are problematic. Hartzell-Nichols finally notes that the debate on intergenerational risks point to the larger problem of anthropocentric versus non-anthropocentric ethics.

Marko Ahteensuu and Per Sandin: The Precautionary Principle

Marko Ahteensuu and Per Sandin discuss the precautionary principle (PP). The PP is often conceived of as a decision-making principle that calls for early measures to avoid and mitigate hazards in the face of uncertainty, in particular, in the context of environmental problems. Ahteensuu and Sandin trace the PP to three sources: (1) the general idea of precaution, (2) nonjudicial codes of conduct and arguments from precaution, (3) and legal documents. They present three ways of conceiving of the PP: as a rule of choice, a procedural requirement, or as an epistemic principle, and the distinction between weak and strong versions of the PP. Ahteensuu and Sandin also discuss a number of common arguments against the PP, such as that it is ill-defined, self-refuting, or counterproductive. They end with observing that formal methods of inquiry have been insufficiently used in the study of the PP. Some topics that also warrant further research are the normative underpinnings of the principle, the status of the principle in risk analysis, and the relationship between the PP and stakeholder/public engagement.

Colleen Murphy and Paolo Gardoni: The Capability Approach in Risk Analysis

Colleen Murphy and Paolo Gardoni discuss the way in which the capability approach might provide for fruitful insights concerning ethical aspects of risk and vice versa. The capability approach has been founded by the economist Amartya Sen and the philosopher Martha Nussbaum and has been extremely influential in the context of development. The capability approach allows to focus on a broader range of capacities, functionings, and achievings than conventional approaches to development that mainly focus on the availability of goods, but not on what people can do with these goods. The authors discuss three ways in which the capability approach can contribute to risk theory: by focusing on capabilities rather than resources or utility, by focusing on threshold levels of capabilities instead of decision procedures, and by focusing on unacceptable or intolerable risks, which avoids the shortcomings of cost–benefit analysis. On the other hand, there are two ways in which risk theory can contribute to the capability approach, by focusing on security as an important dimension of capability, and by allowing a novel way to assess capabilities that looks further than actual functionings achievements.

Part 6: Risk in Society

Given the importance of risk management in modern society, there are several aspects of risk that might, at one point, have seemed to be of mere theoretical interest, but now have vastly important implications for people's lives. The stock and insurance markets rely on methods developed by mathematicians, philosophers, and decision theorists. Increasingly complex technological systems rely on probabilistic methods for safety assessment; methods that could never have been developed without the prior work of risk theorists. In some instances the road from theory to application is short and straight, in other cases long and winding.

As citizens and human beings, we are increasingly required to relate to issues where risk theory is directly relevant in our everyday lives. We are asked to compare insurance policies, invest in the stock market, participate in referendums whether our country should rely on nuclear power or not, and make all kinds of choices where information about likelihood and consequences feed to us from a plethora of different sources.

This has not always been the case. Risk analysis and risk management, as we understand those activities today, are comparatively novel disciplines. We have seen a significant expansion of the field during the last 50 years or so, and the intellectual tools used are modern inventions, where "modern" means at least "post-Renaissance". The most important of these tools—the mathematical analysis of chance events—is in essence a seventeenth-century invention (or perhaps discovery). It was pioneered by thinkers like Pascal, Descartes, and Bayes, and later refined. Pretty soon, it received its applications in the insurance business. The industrial revolution and the increased scope of the consequences of technology—from steam engines to nuclear power plants to genetically modified organisms and climate change—called for rapid development in the field.

Today, risk consciousness permeates nearly every area of societal life. This has not gone unnoticed by risk theorists, and it has given rise to a number of new disciplines, such as the psychology, sociology, and philosophy of risk. The contributions to this section discuss what Ulrich Beck has famously called "risk society," or in other words, the ways society does and should cope with risk.

Rolf Lidskog and Göran Sundqvist: Sociology of Risk

Rolf Lidskog and Göran Sundqvist begin by reviewing the historical background of general sociology. The focus of sociology is the relationship between society and the individual, and this holds also for sociology of risk. The authors sketch the history of the sociology of risk and how it started from experts' recognition that public perceptions of risk differed from those of experts, and the attempts to explain this. They then go on to present three central sociological contributions in which risk occupies a prominent place—those of Mary Douglas, Ulrich Beck (1992), and Niklas Luhmann. Then they present five thematic areas which are subject to intense discussion in contemporary sociology of risk: organizational risk, the relation between experts and public, framing and risk, the epistemological status of risk, and governmentality and risk.

Misse Wester: Risk and Gender: Daredevils and Eco-Angels

Misse Wester notes that empirical risk studies show consistent, systematic differences in risk perception between women and men, with focus on environmental issues and disasters. She identifies three different models to explain these differences that have been proposed in the literature: differences in knowledge of and familiarity with science, biological and social differences, and cultural differences. She argues that each model has problems of its own. She hypothesizes that knowledge plays an inferior role in risk perception in comparison to values, ideology, or cultural belonging and calls for further research in this area. She also calls for further research in the form of critical examination of the function of stereotypes in risk issues, and in the form of investigation of empirical studies of how women and men, respectively, are actually affected by crises and risk on a concrete level.

Tsjalling Swierstra and Hedwig te Molder: Risk and Soft Impacts

Tsjalling Swierstra and Hedwig te Molder discuss a bias in current discourse about impact of technology. Policy makers and experts focus on quantifiable and supposedly value-neutral risk, rather than on other, less obviously measurable impacts, such as emotions, values, and subjective experiences. The authors call this "hard" and "soft" impacts, respectively. They examine how this distinction is theoretically and practically construed, by using two case studies. The first case study concerns an online forum for patients with gluten intolerance, and why some patients reject the idea of a pill that might cure them. The reason is that the pill would affect their identity. A second case study concerns how consumers are concerned about the naturalness of food. This concern is dismissed by experts as a private and invalid preference. The authors analyze how social structures shape the distinction between supposedly valid and invalid forms of and concerns about technological impacts. They distinguish how such impacts are evaluated, estimated, and caused. A better understanding of how the demarcation between hard and soft impacts is construed can contribute to overcoming this bias.

Rinie van Est, Bart Walhout, and Frans Brom: Risk and Technology Assessment

Rinie van Est, Bart Walhout, and Frans Brom explore the relationship between risk assessment (RA) and technology assessment (TA), in particular, parliamentary TA, and how that has evolved over the years since the early days of the Office of Technology Assessment in the USA in the 1970s. The disciplines differ, for instance, with regard to the concept of risk utilized. In RA, risk is typically understood as the product probability and the magnitude of consequences, while TA understands risk in a wider sense, as negative social impact. The authors consider two problems that occur in TA as well as in RA: the problem of representation, that is, who is allowed to define the risk problems under discussion. They discuss how participatory approaches have been developed to alleviate the problem. As a concrete illustration of present- day parliamentary TA, the authors recount the recent TA of nanotechnology carried out by the Rathenau Institute in the Netherlands, and its role in the country's governance of nanotechnology risks.

Marijke A. Hermans, Tessa Fox, and Marjolein B. A. van Asselt: Risk Governance

Marijke A. Hermans, Tessa Fox, and Marjolein B. A. van Asselt write about risk governance, by which they mean attempts at an approach to deal responsibly with public risks which is broader in scope than the traditional categories of risk assessment, risk management, and risk communication, and utilizes several different notions of risk in addition to the classical idea of risk as a function of probability and consequences. They review the origins of the approach and the movement from early positivistic approaches to risk. The term "governance" became prominent in the 1980s, originally in studies of development, and was taken over by other subjects. Today the term "governance," including in risk contexts, is used both in a descriptive and a normative sense, and the distinction is not always clear. Since 2003, considerable efforts have been made by the International Risk Governance Council (IRGC), an independent nonprofit Swiss-based foundation. The authors review current work and analyze it along three lines or principles, which they call "the communication and inclusion principle," "the integration principle," and "the reflection principle."

Marjolein B. A. van Asselt and Ellen Vos: EU Risk Regulation and the Uncertainty Challenge

Marjolein B. A. van Asselt and Ellen Vos introduce what they term "the uncertainty paradox," referring to situations where uncertainty is acknowledged, but where the role of science is seen as providing certainty. The authors argue that it is not recognized that uncertainty undermines the traditional positivist model of knowledge. There are instances of uncertainty intolerance, where uncertainty is not acknowledged and there is unwillingness to produce uncertainty information. To illustrate this, the authors give examples from their analyses of several cases in regulation of risk in the EU, for instance, involving the European Food Safety Authority. They note that uncertainty intolerance is prevalent, but also that there is a tendency to equate uncertainty with risk. They suggest further research involving systematic comparison between risk regulation regimes in different domains. As particularly important topics, they mention the role of science and expertise in decision making and policy, how to deal with uncertainty and trust, the role of the precautionary principle, and stakeholder participation.

Val Dusek: Risk Management in Technocracy

Dusek critically discusses technocratic risk management approaches. The idea underlying much of the current risk management in Western societies is the assumedly superior expert understanding of risk. Such technocratic tendencies in how we deal with risk can be seen in the large number of government committees, commissions, and corporate departments that issue risk assessments and attempt to manage risks. While several contributions to the handbook challenge the quantitative, Bayesian approach to risk assessment, Dusek critically examines the general attitude of mind underlying this approach. Dusek traces technocratic risk management back to the ideal of the superiority of technocratic rationality as advocated by Plato, the seventeenth-century rationalist as well as Francis Bacon and the British empiricists. He explicates and challenges the ideal of an expert rule, following the technology critique of the critical theory of the Frankfurter Schule and existential philosophy, Dusek does not reject risk analysis and risk management, but thinks that the technocratic trend in risk management which builds on the objectivity, universality, and publicity of science, has to be supplemented by other approaches such as the recent work of Gigerenzer.

Conclusion

The Handbook of Risk Theory, of which this book presents a selection, unites scholars from disciplines ranging from mathematics and the natural sciences to the social sciences, humanities, and philosophy. However, as diverse as the approaches and topics are, there is one issue that emerges from practically all contributions, namely that risk involves statistics as much as ethics and social values. There are as yet no final answers on how to deal with risk, nor will there probably ever be such answers, but there nevertheless is a consensus that risk should be approached from different perspectives, including those of stakeholders and the public. This requires "sound science" (broadly conceived, i.e., including social sciences, humanities and philosophy), as much as sound political institutions. When it comes to risk, theory and practice are closely intertwined.

References

Beck U (1992) Risk society: towards a new modernity. Sage, London
Douglas M, Wildavsky A (1982) Risk and culture. University of California Press, Berkeley
Gigerenzer G (2007) Gut feelings: the intelligence of the unconscious. Viking, London
Gilovich T, Griffin D, Kahnemann D (eds) (2002) Intuitive judgment: heuristics and biases. Cambridge University Press, Cambridge
Hume D (1740) A treatise of human nature, 1967th edn. Oxford University Press, Oxford

Ramsey FP (1931) Truth and probability. In: The foundations of mathematics and other logical essays. Routledge and Kegan Paul, London, pp 156–198

Rawls J (1971) A theory of justice. Harvard University Press, Cambridge

Savage LJ (1954) The foundations of statistics. Wiley, New York

Slovic P (2000) The perception of risk. Earthscan, London

von Neumann J, Morgenstern O (1944) Theory of games and economic behavior. Princeton University Press, Princeton

Chapter 2
Levels of Uncertainty

Hauke Riesch

Abstract There exist a variety of different understandings, definitions, and classifications of risk, which can make the resulting landscape of academic literature on the topic seem somewhat disjointed and often confusing. In this chapter, I will introduce a map on how to think about risks, and in particular uncertainty, which is arranged along the different questions of what the different academic disciplines find interesting about risk. This aims to give a more integrated idea of where the different literatures intersect and thus provide some order in our understanding of what risk is and what is interesting about it. One particular dimension will be presented in more detail, answering the question of what exactly we are uncertain about and distinguishing between five different levels of uncertainty. I will argue, through some concrete examples, that concentrating on the objects of uncertainty can give us an appreciation on how different perspectives on a given risk scenario are formed and will use the more general map to show how this perspective intersects with other classifications and analyses of risk.

> I beseech you, in the bowels of Christ, think it possible you may be mistaken (Oliver Cromwell, addressing the Church of Scotland, 1650) (From Carlyle 1871).

Introduction

What we mean by risk is not a clear issue because many writers use the word with slightly different meanings and definitions, even beyond the more vague everyday usage of the term. Aven and Renn (2009), for example, have found ten different

H. Riesch (✉)
School of Social Sciences, Brunel University, London, UK
e-mail: hauke.riesch@brunel.ac.uk

definitions they gathered from the wider risk literature. The problem of clear terminology continues if we go into the various classifications and clarifications of risk and uncertainty, with scholars distinguishing between risks, uncertainties, indeterminacies, ambiguities, and levels, objects or locations of risk and/or uncertainty. With this in mind, I feel slightly apologetic about writing about another scheme devised by myself and David Spiegelhalter (Spiegelhalter 2010; Spiegelhalter et al. 2011), where we use, again, our own terminology, this time in trying to distinguish between different things we can be uncertain about. In this chapter, I will try to explain our distinctions and where they correlate and/or fit in with other classifications of risk and uncertainty, as well as provide an argument on why we feel this particular classification adds to the literature on risk theory by going through a couple of real-world examples.

As Norton et al. (2006) note in their reply to the paper by Walker et al. (2003) discussed below, "an important barrier to achieving a common understanding or interdisciplinary framework is the diversity of meanings associated with terms such as 'uncertainty' and 'ignorance,' both within and between disciplines" (Norton et al. 2006, p. 84). The proliferation of what we mean by risk and how we categorize it within the literature is partly due to the different agendas the different disciplines have with regard to the topic. The question "what do we want to know about risk?" will be answered differently by scholars, for example, interested in risk perception and those interested in the "risk society." Asking this question explicitly may help us in finding out where the different disciplinary approaches to risk intersect. Our classification is partly intended to do just that, mostly because I (as a sociologist) and David Spiegelhalter (as a statistician) have always had slightly different conceptions of what is academically interesting about risk, and our collaboration was partly an attempt to build a conception of risk which is useful for both social and scientific/technical disciplines and will be useful for communicating across this divide by giving a clear account of how and why, in Funtowicz and Ravetz's phrase, there is "a plurality of legitimate perspectives" on risk (Funtowicz and Ravetz 1993, p. 739). At the same time, I hope it will provide a useful and relatively simple map through which the different academic disciplines' interests in risk can be compared and connections seen more easily.

In this chapter, I want to advance the idea that one can confront the different meanings which risk is given and offer an idea of how they are related. There exist many other schemes that try to categorize risks such as Renn and Klinke (2004), Stirling (2007), or van Asselt and Rotmans (2002), and I will try to show how they fit into our overall picture in the following section. As a departing point, I take risk to mean roughly a function of the uncertainty of an outcome and its impact. This definition leaves room for plenty of uncertainties itself, especially since there is no agreement on how to measure impact, or how to compare impacts of completely different categorization, and there is plenty of literature in risk studies devoted to this problem.

The uncertainty part of risk however is itself very problematic. There are some uncertainties we can put a number on, some where we can only evaluate qualitatively and some we have absolutely no idea on how to even start evaluating them. The classification I will propose here is meant to bring some order into the way we think about uncertainty and provide a way in which different types of uncertainty and its classifications can be, if not directly compared, at least brought under the same scheme. Comprehensive surveys of what uncertainties and risks really are and how they should be classified can easily lead to a rather complicated structure that becomes less useful as a heuristic tool for people working within risk. This is more so on the social, policy, and communication aspects than in the technical risk assessment areas, for whom such schemes will be more useful, and I will concentrate on the former in this chapter. Out of the many different dimensions in which uncertainty can be categorized, we chose one in particular which we believe is most helpful when we seek to understand how different people and groups conceptualize and react toward risks. It is meant to analyze risks according to the following question: What kind of thing exactly are we uncertain about?

Background

Philosophical classifications of probability have traditionally focused on questioning where our uncertainty derives from, with the two main choices being uncertainty inherent in the system, and uncertainty arising from our incomplete knowledge. These two interpretations of probability are named by philosophers (Hacking 1975, see also Gillies 2000) *epistemic* probability and *aleatoric* (also often called *ontological* or *ontic*) probability. This basic distinction still underlies modern philosophical theories of probability and can be seen, for example, in the philosophical split between Bayesian (subjective) and frequentist (objective) interpretations of probability in statistics (see also Gillies 2000).

Uncertainty in a larger sense, as opposed to the mathematically defined concept of probability, has also seen attempts at classification. An early and very influential distinction came from Frank Knight, who distinguished uncertainties which are quantifiable which he called risks, and those that are not quantifiable, which he called uncertainties:

> The essential fact is that 'risk' means in some cases a quantity susceptible of measurement, while at other times it is something distinctly not of this character; and there are far-reaching and crucial differences in the bearings of the phenomena depending on which of the two is really present and operating. [...] It will appear that a measurable uncertainty, or 'risk' proper, as we shall use the term, is so far different from an unmeasurable one that it is not in effect an uncertainty at all (Knight 1971 [1921]).

This classification has proved to be very influential especially among sociologists, but is in my opinion slightly unfortunate as it propagates confusion with the traditionally defined concept of risk equaling probability times outcome (or, in the more modern sense focusing on negative outcomes, probability times harm). Although I recognize the usefulness of Knight's distinction for this work, to avoid confusion I prefer to work with the conception that risk refers to a measure of uncertainty combined with the potential outcome.

Combining these two perspectives in a sense, Stirling (2007) recently proposed to divide both the uncertainty as well as the outcome aspects of risk into "problematic" versus "unproblematic" in a similar way to which Knight distinguished between quantifiable and unquantifiable uncertainty. This results in a two by two matrix: at the corner where the probabilities as well as (our knowledge of) the outcomes are unproblematic there are risks associated with the typical statistical risk analyses such as Monte Carlo simulations or costbenefit analyses—these scenarios he terms "risks" in the traditional sense used by most scientists and risk analysts. Scenarios where the probability is knowable, but we are more unsure about the outcomes, he terms "ambiguities"; risks where conversely the outcomes are unproblematic but the probabilities are, he calls "uncertainties." When neither are unproblematic, he talks about conditions of "ignorance." It is worth also pointing out that the term "ambiguity" is used in other disciplines, for example, behavioral economics, to mean unknown probabilities, which is almost precisely the opposite to Stirling's sense—this demonstrates, again, the problems of terminology within the wider risk literature. Technology assessment on the other hand traditionally uses similar terminology but without taking Stirling's ambiguity into account.

Stirling argues that dividing risks into these categories can give us guidance on the circumstances when the precautionary principle could be a valid rule: by dividing risks into qualitatively distinct groups, he argues that the principle can be an important rule for helping with decision making in those circumstances where the outcomes or probabilities are not well understood, and no other type of decision rule would otherwise be helpful.

Another influential attempt at classifying risk elaborated to inform risk assessment policy eventually evolved to inform Funtowicz and Ravetz's very influential concept of postnormal science (Funtowicz and Ravetz 1990, 1993). Funtowicz and Ravetz proposed to map risks as a measure of uncertainty and impact ("decision stakes") and claimed that risks with low uncertainty and impact are the ones familiar from applied science for which traditional mathematical tools of risk analysis are most appropriate. Risks with medium but not high uncertainty and/or impact are in the domain of "professional consultancy," which "uses science; but its problems and hence its solutions and methods, are radically different" (Ravetz 2006, p. 276). The label "postnormal science" applies to situations characterized by high uncertainties and/or high stakes.

Renn and Klinke (2004) similarly use this map with axes denoting uncertainty and impact and identify several areas on that map that delineate qualitatively different risk situations, though these depart from Funtowicz and Ravetz's three areas on the map by being more fine grained: For example, the points in the map where the probability is low but the potential harm is great, they call "Damocles" risks, named after the Greek king who according to the legend had a sword suspended above him by a thin piece of string (the analogy being that the probability of the string breaking at any one point in time is low, but when it happens, the outcome is rather dramatic, at least for Damocles). Points with high probability and high harm they call "Cassandra" risks, after the Trojan prophet who knew about the fate of the city but whose warnings were ignored. Hovering more in the background is a larger area of the map, where we are not very knowledgeable about the event's probabilities or its outcomes ("Pandora" risks).

Brian Wynne introduced his classification of risks as an improvement on the Funtowicz and Ravetz (Wynne 1992) classification which defines postnormal science. Like Stirling, Wynne sees "risks" as situations where the outcomes and the probabilities are well known and quantifiable. Uncertainties are present when "we know the important system parameters, but not the probability distributions" (p. 114). By contrast, the next level, "ignorance," is more difficult to define: "This is not so much a characteristic of knowledge itself as of the linkages between knowledge and commitments based on it" (p. 114). It is "endemic to scientific knowledge" (p. 115), because science has to simplify what it knows in order to work within its own methods. Finally, "indeterminacy" is seen as largely perpendicular to risks and uncertainties, because it questions the assumption on the causal chains and networks themselves. Thus, indeterminacy can be a huge factor in a particular situation even when the risks and uncertainties are judged to be small.

I am sympathetic to Wynne's classification because it recognizes that both quantifiable types of uncertainties as well as the less tangible deeper uncertainties are present at the same time in some situations and thus not mutually independent, which is a necessary realization away from other schemes such as Funtowitcz and Ravetz's map. According to Wynne,

> Ravetz et al. imply that uncertainty exists on an objective scale from small (risk) to large (ignorance), whereas I would see risk, uncertainty, ignorance and indeterminacy as overlaid one on the other, being expressed depending on the scale of the social commitments ('decision stakes') which are bet on the knowledge being correct (Wynne 1992, p. 116).

However, there are for me still some problems with it. First, and more trivially, is the question of terminology. Like almost every other theorist of risk that comes from the social science side, Wynne and Stirling take "risk" itself to be one of their categories, and then proceed to label the other categories somewhat arbitrarily—this results in a mess of technical definitions that leave no special terminology for the overall thing they intend to classify. We cannot call them classifications of risks (or uncertainties) because risk and uncertainty are already part of the classification system. Moreover, this use of the term risk clashes somewhat with the common definition of risk as a measure that combines uncertainty and outcome. This has not helped that another influential tradition of risk theory embodied by Beck (1992) and Giddens' (1999) work takes risk to mean something altogether more nebulous.

Another concern over Wynne's classification, though, is that the categories seem somewhat hard to pin down, in the sense that indeterminacy, for example, includes the various social contingencies that are not usually captured in conventional risk assessments, but what these social contingencies are, and how they relate to the other types of uncertainties is not categorically stated. It is not entirely clear, at least to me, where the boundaries lie, or even if there are supposed to be any precise boundaries. Ignorance, he writes, is "conceptually more elusive" and best explained through a lengthy example. All this in effect makes Wynne's conceptualization hard to explain and therefore possibly inef- fective as a tool for bridging the divide between the social and the technical aspects of risks. The inclusion of broad concepts such as social contingencies as well as quantifiability leaves the feeling that Wynne's categories slice through several useful other distinctions on risk (such as those introduced below, in particular that of Walker et al. 2003). While Wynne's categories are helpful as a conceptual tool to analyze reactions to risk and identifying shortcomings in conventional scientific approaches to risk that need to be addressed, it remains unclear exactly how they intersect and relate to each other. In a way, our own classification presented below is an attempt to reformulate Wynne's insights in a way that makes more intuitive sense and which hopefully helps in addressing the question of how Wynne's categories relate to each other.

Van Asselt and Rotmans (2002) classify risks according to the source of our uncertainty, distinguishing primarily between the two major sources introduced above of epistemic and aleatoric uncertainties (or, in their terminology, uncer- tainties due to lack of knowledge and uncertainties due to the variability of nature). Uncertainties due to lack of knowledge include, for example, lack of observations/ measurements, inexactness or conflicting evidence, while uncertainty due to the variability of nature includes variability in human behavior, value diversity, and the inherent randomness of nature. Aiming to go further than this, Walker et al. (2003) include more dimensions in the classification than merely the source of uncertainty. Thus, they distinguish between location, level, and nature of uncer- tainty: the location uncertainty can be subdivided between context, model, input and parameter uncertainties, and the final outcome uncertainty. Location uncer- tainty therefore roughly describes what we are uncertain about, i.e., "where uncertainty manifests itself within the whole model complex" (p. 9). The levels of uncertainty describe the "progression between determinism and total ignorance" and include, in order, statistical uncertainty, scenario uncertainty, recognized uncertainty, and total ignorance. Finally, the nature of uncertainty is, like in van Asselt and Rotman's classification, mainly about the source of uncertainty, and can roughly be divided into epistemic and ontological uncertainties and subclassified as done by van Asselt and Rotmans.

In this chapter, I hope to be able to add a more inclusive categorization that stays within the spirit of Wynne's as well as Walker et al. (2003) ideas but revolves more centrally around the question of what exactly it is that we are uncertain about, which roughly translates to the "location of uncertainty" dimension in Walker et al. This I will try to use to find interconnections between different literatures on risk. I will argue also that it is useful to apply the scheme to a selection of real-life uncertainties and use it to delineate and make sense of different groups' varying assessments of a situation because they place different importance on the different objects of uncertainty that are all present to various degrees in all of the cases. I will start by making some preliminary distinctions about risk and uncertainty which will enable us to see where this fits into the various other definitions and classifications of risk. I will borrow Walker et al. (2003) idea of different dimensions here, but add that, in our context, these dimensions can best be thought of as different answers to the question on what we want to know about risk.

Firstly, we conceptualize risk as a measure of uncertainty of an event happening times the severity of the outcome. As argued above, this is the usual definition of risk, though it is not used like this by all commentators, some of whom depart more from Knight's (1971) famous distinction between risks as quantifiable uncertainties versus uncertainties that are not quantifiable, which explains Wynne and Stirling's decisions to put "risk" as one of the categories within their overall schemes. Other writers such as those from the "risk society" tradition (Beck 1992 and Giddens 1999) use risk in a much more vague way which is not so much interested in quantifiable or nonquantifiable or even in the separation of uncertainty and severity of the outcome, but sees it more as the vague possibility that things can go wrong. This is again due to the fact that risk sociologists are interested in different aspects of risk (for example, how increasing awareness and preoccupation of risk affects late modern society). There is therefore not much point in criticizing some work for using vague definitions of risk because, from their point of view, there is simply not that much value added to having a precise working definition of what risk is. However, I hope to be able to show how our distinctions can contribute nevertheless to a better understanding of how the conception of risk that is seen as interesting to sociological and cultural approaches can be compared to other conceptions of risk.

Starting from the definition of risk being a measure of uncertainty and severity of outcome, it is secondly to be noted that neither uncertainty nor severity of outcome are in most cases easily measurable or even definable. Our scheme will leave the very interesting problem of severity of outcome for others to work out and concentrate specifically on the uncertainty aspect of risk.

Starting from the question of "what do we want to know about risk?" we can produce a table of different classifications of risk which are designed to answer that question in different ways. We may, for example, be interested in why we are uncertain, we may be interested in who is uncertain, how it affects individuals or society at large, how is risk represented and how should it be represented, and what is it exactly that we are uncertain about? These are the categories I use below, though there will possibly be more dimensions than those, and other authors may want to divide them differently (Walker et al. (2003), for example, distinguish between levels of uncertainty (whether we take a deterministic position or not) and nature of uncertainty (i.e., uncertainty seen as either aleatoric or epistemological), which I would both see as different sources of uncertainty (we can be uncertain *because* we take an epistemological stance and *because* we have an idealized deterministic situation).

Why are we uncertain? Here we can list classifications that have been made regarding the sources of uncertainty, such as in the scheme of van Asselt and Rotmans, which also relies on the philosophical distinction between epistemological and aleatoric (or ontological) uncertainty described above: we can be uncertain either because of our lack of knowledge or because there is an inherent variability in nature. These two fundamental positions are often seen within the more sociological literature on risk as aspects that different situations of uncertainty can take on, so that, depending on the context, an uncertainty can be either epistemological or aleatoric: "it often remains a matter of convenience and judgment linked up to features of the problem under study as well as to the current state of knowledge or ignorance" (Walker et al. 2003, p. 13). In the philosophical literature, by contrast, it is more often assumed that the distinction is a result of different worldviews: we can, for example, be determinists in our general philosophical outlook, in which case, strictly speaking, all uncertainties are epistemic. In most everyday examples, the boundaries of whether an uncertainty should be considered epistemic or aleatoric seems to be a result of the setup, but the precise boundaries or even existence of the boundary to a large extent also depends on our philosophical stances and background assumptions and knowledge. We can, for example, see the probability of winning the lottery jackpot with a given set of numbers as purely aleatoric, because even with the most sophisticated current scientific methods, we are some way away from predicting the numbers drawn even if, philosophically, we are strictly speaking determinists who believe that an all knowing demon could calculate the final result from the initial state. In this example, the existence of probability that is for practical purposes aleatoric even for strict determinists is fairly obvious, though this is not necessarily the case in others. As I will argue below, there are other, epistemic, considerations to be made when we assess the likelihood of winning the lottery.

Who is uncertain? The question of the subject of the uncertainty is interesting from the point of view of psychologists or sociologists, who want to know what effect uncertainty has on people or on society at large. Different people respond to uncertainty differently, as shown, for example, in the well-known "white male effect" and similar phenomena discovered by risk psychology research (Slovic 2000). The subject of the uncertainty is also important for policy making since we would need to know how different groups and individuals respond to risks and representations of risk. For example, my current project investigates local opinions on energy infrastructure: to understand the dynamics of risk opinions within the area the infrastructure is being planned, we need to have a more detailed understanding of who the local actors and groups of actors are and how they interact with respect to interpretations of risk. Whether "the public" consents to the infrastructure being build in their back-yard ultimately depends on a complex interplay between local and national politicians, civil servants, the project developers, media representations, local and national NGOs and residents' interest groups, as well as the individual resident's understanding which is strongly influenced by, and in turn influences, the other stakeholders. Putting "the public" in scare quotes above is meant to signal that there is no monolithic public, with similar agendas, identities, or worldviews. Understanding who the relevant actors are and how they arrive at their conceptions of risks and how they influence and are influenced by other groups of actors is vital for the analysis of what role risk plays in planning decisions.

How is uncertainty represented? Representations of uncertainty can take on different forms, which is again related to where the risk stands and is perceived along the other dimensions. We can, for example, simply deny that there is any uncertainty or risk at all or just concede that there is some, but more or less, undefined uncertainty. If we want (and know more about the situation), we can give a list of possible outcomes, either on their own or with some indication, qualitative or quantitative, on how likely each outcome would be. Should we have chosen a model we think is appropriate, we can give the result of the risk assessment as say a probability, with or without error bars or other representations of uncertainty on that final number.

How we represent risks depends very much on our knowledge of the situation, denying risk is a valid action when we do not know of any, and a simple list of possible outcomes is useful when we lack knowledge of how likely each outcome would be. However, which representation people chose in practice often depends also on what message they want to get across, or even reflects philosophical stances or implicit assumptions made. For example, if we want to make the risk of taking a particular medication look high, we can choose to represent it in relative rather than absolute terms. Similarly, we can give a positive or negative frame: for example, there is technically no difference between saying that "your chance of experiencing a heart attack or stroke in 10 years without statins is 10 %, which is reduced to 8 % with statins" and "your chance of avoiding a heart attack or stroke in 10 years without statins is 90 %, which is increased to 92 % with statins"—yet these two formulations have different connotations for the reader (example taken from Spiegelhalter and Pearson 2008).We can express probabilities in percentages or "natural frequencies," where research has shown that people are intuitively better able to understand natural frequencies (Gigerenzer 2002). We can produce bar charts, pie charts, "smiley charts" on top of the verbal expressions, and these again convey different impressions of how risky something is. Finally, we can express uncertainties according to our philosophical understanding—if we say that I have a 10 % chance of having a heart attack within the next 10 years, that can either mean "10 % of people with test results like me will have a heart attack," or "10 % of alternative future worlds will include me having a heart attack." Again these scenarios while both expressions of the same amount of uncertainty will qualitatively feel different to people, with the second usually seen as the more persuasive way to get people taking their medicines, because it is more personalized (see also Edwards et al. 2001 on the effects of framing risk to patients).

Responses to uncertainty: How do people react to uncertainty? Do we, or should we, respond rationally to risk, for example, by doing a cost-benefit analysis to evaluate risks (Sunstein 2005)? Slovic et al. (2004) argue that while analytical and affective are two distinctive ways of reacting to risk, they interplay to produce rational behavior. But maybe even this distinction between affective and analytical needs to be challenged (Roeser 2009, 2010).

On a larger societal level, the risk society literature concerns itself, among of course other things, with how a society responds to risks (specifically our own, late modern—i.e., contemporary Western-society). Here, the issue is not so much about the nature of the risk as such (though it plays a role as I will outline below), or even whether the risks are real or not, but with the role that an increasing awareness of risk plays within late modern society. In particular, they describe the intuitive pessimistic induction through which people have come to realize (or at least believe) that there are always unexpected uncertainties and the possibility of things going horribly wrong with any possible new technological invention (the "unintended consequences of modernity"). Thus, as society has become more reflexive about its own technological achievements, the awareness of risk has become a more powerful driver of social forces than it was previously when risks were more perceived as due to intangible forces of nature rather than consequences of our own society, and therefore modern Western society's response to risk has become qualitatively different to what it was before.

Understanding uncertainty: How people understand uncertainty is a related but somewhat orthogonal issue to the above—this may relate, for example, to the literature of social representations of risk (Joffe 1999; Washer 2004), which uses the social psychological literature of social representations (Moscovici 2000) to characterize how risk issues are perceived and made sense of through associated reasoning—new abstract and intangible concepts as are usually found in topics surrounding risk are conceptually anchored to concepts that are already under-stood, and thus new concepts are better assimilated into a group's already held worldviews. Washer, for example, describes through the analysis of newspaper reports of recent new infectious disease outbreaks like SARS or avian flu and how these unfamiliar diseases (and the risks they represent) are being commonly anchored to already understood and familiar diseases (erg. the Spanish influenza outbreak of 1918), or to other aspects, such as vaguely xenophobic expectations of lax health and hygiene practices of the countries of origin; these mechanisms thus place the new disease into different categories of risk than they might otherwise have been perceived if anchored differently.

Hogg (2007) similarly uses a social psychological perspective, social identity theory (Tajfel 1981) to describe issues of intergroup and in group trust, arguing that our social identities about which groups we belong to effect how we trust the risk statements of others—in-group members are trusted more than out-group members, and even within groups, individuals who are more prototypical in that their characteristics conform well to group norms and values, are trusted more than more marginal members.

Here, we can also list other approaches that are interested in the social construction of risk. Cultural theories of risk such as the influential approach of Mary Douglas (1992) and the risk society argument are relevant here as well, because it is concerned with how societies construct (and thus understand) risks.

What exactly are we uncertain about? I left this category until last because this will be my focus in the following section. The object of our uncertainty has been the concern of several classification systems described above when, for example,

Wynne talks about uncertainty over causal chains or networks. Similarly,Walker et al. (2003) talk about "locations" of uncertainty, defining that as "where uncertainty manifests itself within the whole model complex" (p. 9), and distinguishing between uncertainty about the context (uncertainties outside of the model), models, inputs, parameters, and the final model outcome. In the following section, I will propose our slightly similar scheme which aims more at a rather general classification of the main types of objects we can be uncertain about which translates somewhat into Walker et al. locations of uncertainty, but is aimed at dispensing a too fine-grained classification in favor of one that we feel makes intuitive sense and can help explain different groups' reactions toward the same risk scenarios.

In slicing the risk literature into these different categories of what they find interesting about risk, I recognize that a lot of work on risk looks at interactions between these different categories: for example, we can be interested in how different representations of risk and different aspects of risk can affect different people or groups of people. But I hope that this way of presenting the risk literature helps make sense of these interactions, and can therefore provide an interesting look into how different aspects of research on risk interlock. Our specific distinction between different objects of risk is itself designed in part to explain different outlooks on risk. In the following sections, I will present the different objects of uncertainty and, following that, explain through a few examples of how objects of uncertainty interconnect with some of the other dimensions of risk in a way which will hopefully give us a fuller description of the different risk scenarios.

Objects of Uncertainty

Our classification (Spiegelhalter et al. 2011), somewhat unwisely in retrospect, divides the objects of uncertainty into different "levels." I am calling our decision unwise because this suggests a particular linear hierarchy which may be misleading, but also because other commentators have attached the label "level of uncertainty" to some of the other dimensions of uncertainty outlined above. Specifically, Walker et al. use the term "levels of uncertainty" to describe the spectrum from determinism to "total ignorance."

We distinguish between three types of uncertainty within the modeling process, and two without. Our use of the term model here is meant to be rather generic. Philosophical and social studies of scientists have shown that the term "model" can be used in varying ways in science (Bailer-Jones 2003), and, thus, generally it does not have the precise definition that it would have in mathematics or statistics. For example, the everyday constructions through which we as laypeople make sense of risk situations is taken here to be a kind of model as well, since we take our own incomplete information of the world and how we understand things to work and thus gain an understanding of what might happen. The difference between the nonexpert modeling we do in our everyday life and the expert risk assessments is at the end merely a matter of background technical knowledge and competence and levels of commitment, rather than a huge qualitative difference. There is of course much more to be said about lay understanding and construction of risk perspectives, but, for my purposes, it should be enough to use the term "modeling" in an inclusive way that encompasses both expert and lay processing of risk.

By using the term model in this broad sense, I can apply it to different and varying real-life uncertainties and can include the formal mathematical application of the term as well as the more vague, everyday usage of model, in order to achieve applicability of our scheme across a wide variety of real-life risk situations, where precise mathematical or statistical modeling is impossible, impractical, or simply overlooked.

The categories I will present here are not meant to be mutually inclusive, and they will overlap. On the contrary, as I will argue with a couple of examples, in most risk situations, various levels of uncertainty are present at the same time, and our differences of opinion about risks may be due to us giving different importance to different levels.

Level 1. Uncertainty about the outcome. The model is known, the parameters are known, and it predicts a certain outcome with a probability p. An example here is the throw of a pair of dice: Our model is in this case the fundamental laws of classical probability, the parameters are the assumption that the dice are fair and unloaded, and the predicted outcome of, say, two sixes is $(1/6)$ $(1/6) = 1/36$.

This is comparable to the "final model outcome" in Walker et al. On its own, this level of uncertainty exists only in rather idealized situations, as in arguably the example above of the dice. However, this is the level at which we as members of the public are most likely to encounter risk, for example, when we read in a newspaper that "the chances of developing bowel cancer is heightened by 20 % if we eat a bacon sandwich every day" (which is a real example taken from the case study further elaborated below). Such clear numbers, in the vast majority of cases, hide the fact that there are additional uncertainties related to the process in which experts arrived at it.

Level 2. Uncertainty about the parameters: The model is known, but its parameters are not known (Once the parameters are fixed, then the model predicts an outcome with probability p).

This may simply be a lack of empirical information: If only we knew more, we could fix the parameters.

Our concept of uncertainty about the parameters itself hides a variety of different ways in which we can be uncertain about them: We can have fairly good, quantified probabilities about what the parameters should be as they might simply be a matter of getting better information about the system that is being modeled, but, more problematically, we could also be uncertain about how better measurements themselves are achieved, and/or our uncertainty about the parameters can itself only be expressible as a probability distribution, or even only a qualitative list of possibilities, or lastly we might simply have no idea of what the possibilities could be in the first place. Thus, here some of the different dimensions as outlined above intersect with the object of uncertainty: our uncertainty about the parameters can be due to epistemic or aleatoric sources, and it can be represented in different ways.

Unlike Walker et al. we make no distinction between parameter and input uncertainty here, firstly for reasons of simplicity, but also because this more fine-grained distinction is not all that useful when we try to apply our scheme to real-life examples. Similarly, we would also class uncertainties over boundary conditions and initial values into this category as well, all of which may strain the term parameter uncertainty into categories not strictly speaking considered parameters as such—at the end, however, we decided to balance usefulness and simplicity with detail.

Level 3. Uncertainty about the model: There are several models to choose from, and we have an idea of how likely each competing model is to reflect reality. Models are usually simplifications about how the world works, and there are often several ways of modeling any given situation.

This is analogous with Walker's model uncertainty, and again, this uncertainty itself can be presented in different ways and may be due to different sources. The way we should represent the uncertainties over model choice is more of a contentious issue and, of course, depends on the precise source of that uncertainty itself. In Spiegelhalter et al. (2011), we advocate a Bayesian approach to compare competing models (after Hoeting et al. 1999). This though will not be everyone's favored approach, which means that, in most situations, we will encounter varying approaches to representations of model uncertainty.

Here, there can be, and frequently are, disagreements between the experts themselves, which means that to the nonexpert public or other consumers of a risk assessment, the uncertainty over the model choice is often related to other factors, such as how much trust they place in the experts to evaluate their model choices honestly or competently, and involves furthermore making a judgment between different experts' assessment when faced with disagreement—however competence and honesty are assumptions that are made only implicitly (only rarely will experts be honest enough to consider their own competence as part of the overall risk assessment—building in an estimation of your own honesty into a risk assessment poses even more problems) and not strictly speaking part of the modeling process. There is therefore is a qualitatively different uncertainty for consumers and for producers of risk assessment, which will be the next level.

Level 4. Uncertainty about acknowledged inadequacies and our implicitly made assumptions. Every model is only a model of the real world and never completely represents the real world as such. There are therefore inevitable limitations to even the best models. These limitations could arise because some aspects that we know of have been omitted, or because of extrapolations from data or limitations in the computations, or a host of other possible reasons. Similar in a way to Wynne's concept of "indeterminacy," this is about questioning the assumptions we make, for example, about the validity of the science itself, and thus goes slightly perpendicular to the problems of choosing the models and parameters. These include the "imaginable surprises" (Schneider et al. 1998), that is, things we suspect could occur but about which we do not know enough to be able to include them in the model.

As outlined above, this is where the question of trust comes into force as these are factors that are implicitly not assumed to matter in the risk assessments but not (or rarely) part of it. Similarly, there are always many assumptions about the world that have to be made and that are not part of the modeling process because they are assumed for one reason or another. For example, the risk referred to above of eating too many bacon sandwiches relies not only on the empirical and theoretical studies performed in the analysis, but also on the accumulated medical knowledge about cancer that was taken as given within the risk analysis. Any error within the fundamental scientific background assumed in a model that is supposed to reflect the real world albeit simplified will make that model less reliable. Therefore, uncertainties in our assumptions and scientific background knowledge are also inadequacies that are acknowledged but not usually part of the modeling process itself. At the same time, these are inadequacies in the process that are at least acknowledged in some way even if not particularly acted upon.

Dealing with acknowledged inadequacies can be done through informal, qualitatively formulated acknowledgment or listing the factors that have been left out of the model, or of course simple denial that there are any in the first place.

Level 5. Uncertainty about unknown inadequacies: We do not even know what we don't know. This particular type of uncertainty was made notorious through Donald Rumsfeld's famous speech on "unknown unknowns":

> There are known knowns. These are things we know that we know. There are known unknowns. That is to say, there are things that we now know we don't know. But there are also unknown unknowns. These are things we do not know we don't know (Rumsfeld 2002).

There are as yet not very many formal approaches to unknown unknowns literature, though the concept has been well known for a while—for example, Keynes wrote that about some uncertainties, "there is no scientific basis on which to form any calculable probability whatever. We simply do not know" (Keynes 1937). Long before Keynes and Rumsfeld, however, a concept similar to unknown unknowns was introduced by Plato through the famous "Meno's paradox": How can we get to know about something when we are ignorant of what it is in the first place? (Sorensen 2009). It is also related to Taleb's concept of "black swan events" in economics (Taleb 2007) which are events that were not even considered but which, due to their high impact, have a tendency to completely change the playing field, and which is one of the concepts he used to warn about (what turned out to be) the 2008 world financial crisis.

These inadequacies are difficult to deal with formally or informally because we don't really know what they may be, and we are constrained in a way by the limits of our imagination of what could possibly go wrong—Jasanoff (2003), for example, identifies lack of imagination as one of the factors limiting our knowledge for proper risk assessments in postnormal science (p. 234).

Responding to unknown unknowns is naturally very difficult because by definition we do not know what they are. We can however acknowledge them through simple humility that it is always possible that we are mistaken, as demonstrated by Cromwell's quote in the epigraph. Another way is to brainstorm every possibility we can think of and letting our imaginations go wild. This approach is of course never going to be able to cover everything that could go wrong and will therefore not eliminate unknown unknowns.

A slightly more formal way of responding to unforeseen events is the introduction of "fudge factors," for example, in bridge or airplane design, where we design the structure to be a bit stronger than even the worst case scenarios that we could think of require—though even then there is always the conceivable possibility that something worse may happen.

These levels in a way relate to different concerns of different disciplines—who are after all interested in different aspects of risk. For example, the traditional mathematical and philosophical problems of probability theory are mostly concerned with level 1 uncertainty. Statisticians are mostly concerned about level 2 and 3 uncertainty, that is, finding the right model and, within that model, adjusting the parameters appropriately. It seems unfortunately that uncertainties on which we cannot have a particular mathematical handle on are so often ignored by statisticians and risk modelers—often probably for the pragmatic reason that there simply is not much they can say about the higher levels with the mathematical tools of their trade. Shackley and Wynne (1996), for example, write that in their study of policy discourse on climate change, policy makers were concerned about the validity of the models, while the scientists themselves never even considered that to be an issue, but were instead more concerned about measurement errors within their models. This is to an extent an unfair generalization. An informal survey of technical abstracts from a recent Carbon Capture and Storage conference (Riesch and Reiner 2010) has shown that while model uncertainty is not generally discussed, it does occasionally get mentioned, alongside even an occasional awareness that there are uncertainties associated with unmodeled or unmodelable inadequacies. Nevertheless, worries about model inadequacies were certainly not a prevalent concern among the scientists and risk modelers.

This expert discourse unfortunately distorts the way we perceive particular risks because higher level uncertainties still exist. This may lead to situations like the ones described by Taleb (2007) when he writes about economists having forgotten that unforeseen out-of-the-blue events can occasionally happen and completely mess up our predictions—the sort of events he calls "black swans" if they also have a high potential impact.

The risk society approach of Beck and Giddens is talking mostly about levels 4 and 5, where it is hypothesized that late modern society is living with the increased realization that unmodeled and unmodelable risks are pervasive, and that even if we had some kind of handle on them, there is always the possibility of completely unforeseen events, what Beck calls the "unintended consequences" that he mostly associates with new technology, but which need not necessarily be tied in with it. In Beck's characterization of late modern society, we have now become accustomed to the realization that despite the best risk modeling of science and engineering experts, technological innovations and advances always have unforeseen consequences, completely left-field occurrences that the original evaluations failed to take into account—in other words, we now know that we live with level 4 and 5 uncertainties all around us. In a way, it matters less to the sociological literatures whether these risks are real or not, but the mere realization that they do happen affects the way late modern society evaluates technological progress and ultimately, itself. Beck's work has been criticized for ostensibly being about risk, but not quite understanding the concepts of risk analysis and probability (Campbell and Currie 2006), though this slightly misses the point because, within this scheme, it is not really the nature of risk that is important, but responses to it.

As I have tried to argue above, the different disciplinary approaches to risk intersect in different ways—not only do they find different objects of uncertainty important, but they are also interested in different topics among the other dimensions. However, I have not yet found a comprehensive way of translating between the different approaches, and my categorization between different dimensions of risk is meant to solve this. In particular, I feel that the objects of uncertainty dimension which I presented here in more detail can be an important perspective with which to analyze different risk situations in a way that makes sense to the different disciplinary approaches. In the following section, I will go through several examples to illustrate what this perspective can show how all levels of uncertainty are present in most situations involving risks or uncertainties. In particular, I am interested (as a sociologist) to explain how different groups' perceptions of essentially the same scenario can differ so dramatically: because through their background experience, assumptions, and worldviews, they will attach different importance to the different levels described above. One important departing point therefore is my assertion that all the levels are present in every risk situation, and that the relative importance that is attached to them depends on who is mulling over it, and I will argue for that below. This seems to be more important on the objects of uncertainty dimensions more than on some of the others, and therefore I feel concentrating on these will help us bring about a more comprehensive way of translating between the various risk literatures, as these will be interested in different objects of uncertainty within each situation.

Examples

In this section, I will explore how these five levels of uncertainty can help explain what happens in various real-life cases in which risk, perceived or real, is a factor, and how our concept of the levels explain different perceptions and how this can lead to the communication difficulties between groups with different perspectives (say between to proponents and opponents of carbon capture in the third example). In each of these cases, all five levels of uncertainty are present, though they are differently important and relevant depending on the example.

The Lottery

I will start with a situation which traditionally is seen as less problematic because it seems to rely only on outcome uncertainty. In a typical national lottery, such as in the UK, there are 49 balls, and each week, six balls are drawn; people who have chosen all six correctly win the jackpot. While the exact rules of how much you win are more complicated (depending on the lottery), the case is at least on the surface clearly of level 1: The model is known, the parameters are known, there are no known inadequacies in the model; all the uncertainty that remains is the probability predicted by the model.

This however does not mean that there are no uncertainties present of the other levels, they are simply more hidden and seem less relevant. Level 2 uncertainty concerns the uncertainty of the parameters. In this case, one of the parameters that we have assumed were fixed concerned the individual probability for each ball to come up, thus the question essentially revolves around whether the lottery machine is fair. This is of course a question that we should be asking ourselves when we play the lottery, though we rarely do because we trust the authorities that set up the game. As soon as that trust is lost however, level 2 uncertainty comes to the foreground in our evaluations of how likely a jackpot win is. But this is also an empirical question—for the regulator to make sure that the parameters are what we assume them to be, the equipment is regularly checked, and therefore even if we trust the operators to run a fair game, there is still residual empirical uncertainty over the measurements performed during the equipment checks.

Level 3 and 4 uncertainties, in this case, are less likely to bother us because the situation is relatively simple. We, thus, do not really have competing models with which to describe the lottery: unlike in the examples below, where we have situations for which we need a model to describe it, in this case, we start with the model and set up the reality to fit it—that is, after all how the game was constructed. Therefore, in this case, we have a lot of confidence that the mathematical model we use to describe the game is accurate and not likely to be replaced by one that reflects the situation better.

This however does not necessarily reflect the situation from the point of view of the consumer of the lottery—given that the rules of the game are published but the precise probabilities for a given type of win are not necessarily, we have to make our own calculations, and, for the mathematically less able among us (such as myself), there remains the very real possibility that I have made a mistake in estimating my chances of winning. Again the lottery is a pretty simple situation where even I will not have many difficulties; however, the same cannot be said about other games of chance such as blackjack where there is no real model uncertainty from the point of view of an able calculator, but where for the average player, the probabilities are very much subject to uncertainties over mathematical ability. The trust that we have in the operator mentioned above to demonstrate why our parameter assumptions may be wrong is itself a frequently *unacknowledged* inadequacy: The probability that the operator is cheating is, even if somehow quantifiable, rarely part of the model (which in turn makes out the parameter uncertainty to be only dependent on empirical questions) used to estimate the probabilities of winning or the expected pay-out. Yet again, for the consumer who may have a different estimation of the trustworthiness of the operator, level 4 introduces an unmodeled uncertainty, and their estimations of this uncertainty will be different according to background knowledge and assumptions.

Considering completely unexpected scenarios now, maybe the machine could blow up during the draw, invoking maybe the need to refund punters—again this would affect the probability of winning overall in a slight way. Or the operator could be declared bankrupt, in which case, it is not clear whether there would be refunds at all, and the issue would probably only be solved on a case by case basis depending on the whims or political pressure of put upon the government as is the case when other companies fold (even though costumers will usually not get refunds if a company goes bankrupt, there are often cases, such as tour operators, where political intervention may make an exception). The possibilities here are of course only restrained by my imagination and as soon as I formulate them they are not strictly speaking unknown unknowns. However, the relative ease with which we can conjure up scenarios which are not foreseen at all points toward a large background level 5 uncertainty which cannot be eliminated completely or even adequately estimated through better modeling.

These considerations I hope demonstrate that even in seemingly very clear situations that are not usually assumed to be subject to other than level 1 uncertainties, our estimation of the uncertainties rely to a large extent on our trust in the operator, our background assumptions and mathematical abilities, and these differ from person to person.

Saving Our Bacon

What exactly does it mean when we are informed that we are facing an increased risk (by 20 %) of bowel cancer if we eat more than 500 g of processed meat a day (WCRF 2007a)? Again, I will hope to demonstrate here that in this claim, there are several levels of uncertainty interwoven because, depending on which perspective we take, we can evaluate the uncertainties of different objects in varying ways. Therefore, making sense of that claim will involve untangling them. (Incidentally, the lifetime risk of bowel cancer is estimated in the report as 5 %, which raises to 6 % when we eat a lot of red meat a day. In relative terms, the increase of risk is 20 %, while in absolute terms it is 1 %. The fact that the WCRF chose to present the more scary relative increase in their press strategy, rather than the more informative absolute increase, tells us a lot about their communication priorities, see also Riesch and Spiegelhalter 2011).

The claim above is based on a meta-analysis performed by the World Cancer Research Foundation of various published trials that investigated the incidence of bowel cancer among people who consume a lot of red meat versus those that do not (WCRF 2007a). One level of uncertainty therefore involves what the studies, as aggregated by the accepted rules of how researchers should do meta-analyses, tell us about eating processed meat: the model is known (in this case, the rules involved of doing the analysis, as well as the rules of the individual studies aggregated in the meta-analysis), the parameters are fixed (in this case the empirical evidence), and together they predict the outcome, bowel cancer, as 20 % higher than without the consumption of processed meat. This is the level of uncertainty at which the WCRF communication strategy operated: Our science has found that the risk is p, and that is what the public should know about red meat (as suggested by the WCRF press strategy; WCRF 2007b; 2007c).

However, especially when looked at from the perspective of the reader of the report, the other levels of uncertainty are there in the background as well and have been emphasized by some of the other actors in the debate: Level 2 uncertainty is, to a certain extent found in the report itself, as this represents the empirical uncertainties surrounding each individual study in the meta-analysis (i.e., fixing the parameters through empirical data): These empirical errors have been aggregated, and since the meta-analysis involved lots of different studies, the overall error has been reduced, and this level of uncertainty is represented through the use of error bars in their charts. Error bars of course did not make it into the verbal communication that accompanies the study's conclusion; instead, the information about uncertainty here is formulated qualitatively: The report distinguishes between the evidence being "convincing," "probable," and "limited." In the final communication of the report, the inherent experimental error, the level 2 uncertainty, was not quantitatively included and could possibly be said to be relatively low. There were though a small qualitative indicators in the wording of the press release:

> *There is strong evidence* that red and processed meats are causes of bowel cancer, and that there is no amount of processed meat that can be confidently shown not to increase risk (WCRF 2007c, my emphasis).

While level 2 uncertainty has been addressed in the report, if only qualitatively as a way of showing some caution in interpreting the results, level 3 uncertainty posed more problems and was the sort of uncertainty that the expert critics of the report have focused on: This is uncertainty surrounding choosing the model itself. In this case, that translates to the controversy of how the meta-analysis was done, and specifically which studies were included in the analysis. Critics of the report have pointed out that the meta-analysis has left out many individual studies that, if included, would have given the whole analysis a different result. Whether or not there was much merit in these criticisms, they at least demonstrated that no amount of certainty in the analysis itself can remove the uncertainty inherent in choosing the model. The methodology of meta-analysis in general, while an established tool within medical research, is nonetheless not without its critics, and again, though I will not comment on whether these criticisms have much merit, they demonstrate that even within the expert community there are differences of opinion, and therefore, especially for nonexpert bystanders like me, there is an additional uncertainty over whether the whole methodology used by the report is sound in the first place.

    This is compounded by level 4 uncertainty because even if we did have certainty over which studies should have been included, and whether meta-analysis in general is the best way to pool the results from these studies, there is still residual uncertainty about the scientific background assumptions underlying the study, which relies to a large part on previous medical knowledge on for example cancer which is seen as well-established and therefore not considered a factor to be included in the model at all. This is to an extent not too much of an inadequacy because the study being an empirical evaluation of several selected trials and observational studies does not rely much on previous medical knowledge; however, it does rely on previously established medical and scientific knowledge and assumptions that these methodologies are a valid way of establishing knowledge. Again though it is not my place to comment on whether there is any merit to these criticisms, that is a criticism that certainly has been made, not so much by medical experts, but by alternative health practitioners who reject a large amount of otherwise established medical knowledge and methods. For the nonexpert bystander, again, the situation is that of competing groups who both claim to have expert status and who have different opinions about what the study shows and can even in principal be expected to show. This is then a different level of uncertainty altogether for the consumer of the report.

    Added to that, there are other implicitly made assumptions in the report which relate to the honesty and competence of the researchers themselves. Of course, we can't expect them to take these concerns seriously as an additional uncertainty in their own science, but these are not assumptions that can automatically be assumed by the reader. Both of these two levels of uncertainty (3 and 4) were emphatically not voiced in the official communication by the WCRF, which is understandable because they would have cast doubt on their own experts' judgment. However, they were certainly voiced by the critics: Level 3 uncertainty, as shown above, was the expert critic's response, of fellow medical researchers who accept the general methodology but object to the way it was performed in this instance, while level 4 uncertainty is more usually the response of critics who disagree with meta-analyses generally or who distrust or disbelieve some of the assumptions that medical research takes as established (these are not very influential among medical researchers, but have some influence among alternative medicine campaigners).

    Finally, there is level 5 uncertainty: We may be completely wrong footed about the risks of processed meat—maybe the results of all the studies were a systematic error in the design of contemporary medical studies that we do not know about? Maybe something even more exotic has gone wrong? Admittedly, in this particular case, it is quite hard to imagine possible level 5 uncertainties, but this is of course the nature of this level of uncertainty as I have defined it by us not having any handle on it, and not even ever having thought of the possibility. Accordingly, giving a numerical estimation of this level of uncertainty is impossible. Beyond gut instincts, we cannot even tell if it is likely or not likely that something is fundamentally wrong with our conceptions of the problem. In the next examples, I will show that while in this case level 5 uncertainty is not much incorporated in the current thinking about the subject, in many other situations involving risk, level 5 uncertainty can be central.

The complications and disputes involved in this case are connected with the protagonists talking about different levels of uncertainty: The WCRF experts talked about level 1 uncertainties in their take-home message to the general public, while at least acknowledging level 2 uncertainties when talking among themselves and in communication with other experts. Level 3 uncertainty is the level at which the expert critics attack the report, while the less listened to nonmedical critics attacked it at level 4. Meanwhile, level 5 uncertainty looms menacingly in the background. Some of the media interpretation and discourse about the WCRF study and its press releases can be found in Riesch and Spiegelhalter (2011).

Carbon Capture and Storage

Carbon Capture and Storage (also referred to as Carbon Capture and Sequestration), or CCS, is a technology designed to reduce carbon emissions from fossil fuel burning power plants by capturing the CO_2 through various processes and storing it underground in depleted natural gas reservoirs or other suitable storage sites. The technology is seen by its proponents as an important and technically feasible way to lower carbon emissions because it relies on already fairly well-known mechanisms. While it is admittedly only planned as a relatively short-term solution to be deployed while renewable energy sources are being developed further, it solves some of the problems of "technology lock-in" that could happen if we concentrated only on a few favored energy sources such as solar- and wind-power which we have no guarantee yet that they will be deployable at a large enough scale to reduce carbon emissions in time to avert catastrophic climate change. Therefore, it is seen as part of a necessary portfolio of energy technologies that needs to be included if we want to avoid putting all our eggs in one basket (as argued for example by the influential Stern report, Stern 2007). Further benefits of the technology include more security in the energy supply because it would make burning coal an environmentally sound energy option again and, therefore, reduce the dependency some countries with large coal resources like the UK have on foreign gas imports. A more environmentally appealing further benefit of CCS is that when the technology is developed far enough, it can be used in conjunction with biomass burning power plants and therefore represents one of the few currently technically feasible ways of removing carbon from the atmosphere.

Despite these advantages, CCS has many opponents principally among the environmental community, who argue that it merely propagates our dependency on fossil fuels and drives funds away from developing more promising energy technologies which need to be developed anyway because even proponents of CCS see it only as a short term solution (Greenpeace 2008). Finally, one objection to CCS which threatens to be a show-stopper is the safety risks to local people and the local environment that are posed by possible CO_2 leakage from the storage reservoirs and the pipelines that transport the CO_2 from the power plants to these sites. It is these safety risks of CCS that I will concentrate on here; however, the other arguments for and against CCS are relevant here because it is our background worldviews, knowledge, and assumptions which color the way we perceive specific risks. One immediately obvious example of how our background knowledge may color our perception of the risks of the technology concerns our knowledge (and uncertainties within that knowledge) of the toxicity of CO_2. While CO_2 is not, in fact, neither toxic nor flammable (it does, however, act as an asphyxiant, and therefore still represents a potential though somewhat lessened danger to people living near leakage sites), public opinion surveys on perceptions of CCS have shown that worries over CO_2 are very much in the forefront of public safety concerns (Itaoka et al. 2004; Mander et al. 2010).

Level 1 uncertainty in this case is the final number of the risk assessment, which is usually the basis on which politicians or energy companies would claim that experts find the technology to be very low risk. These numbers are arrived at through models which make of course several assumptions. A general model of how carbon storage works depends very much on local conditions if we want to arrive at numbers for any particular reservoir and the surrounding area. The local conditions vary in great detail, and therefore experts who perform risk assessments of prospective sites need to investigate them very closely—what are the exact geological formations the CO_2 would be stored in, what are the properties of the cap-rock formations that are needed to keep the CO_2 from traveling up, are there seismic fault lines and if so how would they affect the storage, how many man-made injection wells are there, and how exactly are they going to be sealed once the CO_2 is injected, what is the general threedimensional shape of the landscape above the possible leakage sites (since CO_2 is heavier than air, there is a chance that it might stay if it leaks into a valley and thus cause greater potential health risks). All these are in a sense parameters that need to be put into the general models if we want to arrive at a final number of expected deaths per year. All of these are subject to their own uncertainties, either because of potential measurement error, or even a more general lack of understanding of the local conditions which in practical terms can only be estimated.

At level 3, there is the choice of general model. In the case of CCS, there is still some argument over whether models developed by the gas and oil industries are really applicable to the storage of CO_2 (Raza 2009). There are also potential debates to be had over precisely what statistical methods should be used and their applicability. Writing in a Dutch popular science magazine article about the proposed (now canceled) CCS storage site under the town of Barendrecht near Rotterdam, Arnoud Jaspers felt that there is some additional uncertainty over model choice when he interviewed modeling experts, for example, some of the models simply did not take into account the three-dimensional structure of Barendrecht and therefore arrived at unreliable scenarios of what would happen should CO_2 leak (Jaspers 2009, 2010). As this shows, not every relevant bit of information makes it into all the models, and there is therefore some uncertainty about which model would be the best to use.

Furthermore, there are other things that are not considered in any model because we simply do not know enough about them or their relevance to be able to model them, these then are the acknowledged inadequacies we term level 4 uncertainties. A recent draft guidance document by the European Union on the implementation of Directive 2009/31/EC concerning geological storage of CO_2 (EU 2010), for example, divides the types of risk expected from reservoir leakage to be "Geological leakage pathways," "leakage pathways associated with man-made systems and features (i.e., wells and mining activities)," and "other risks such as the mobilization of other gases and fluids by CO_2)" (p. 31). While the first two are routinely part of the models, the "other" category provides more of a problem because, other than listing some possible scenarios that can only to be considered on a "case by case" basis, there is not that much additional analysis that can be introduced. Therefore (as a brief glance through the technical papers on CO_2 storage at the GHGT10 conference has shown—discussed in more detail in Riesch and Reiner 2010), most risk models of CO_2 storage consider leakage pathways along geological fractures or man-made boreholes but do not as such feature other possible leakage pathways either because they are judged to be not very important or, more worryingly, not enough is known about them to include them in the models.

Lastly, there are level 5 unknown unknowns which are not part of the modeling process, not because we do not know enough about them, but because we do not know about them at all. Giving a concrete example is, again, impossible since simply by thinking about them they become known unknowns. However, there are scenarios that can be imagined by the public that are never even considered in the expert literature. For example, the cover illustration to Jaspers' (2010) popular science article on CCS features a huge "blow-out" scenario with a vast amount of CO_2 escaping explosively and destroying large parts of Barendrecht's neighborhood with rescue helicopters hovering around the scene like tiny flies in relation to the explosion; painting the picture of carbon storage as a huge shaken soda bottle bubbling menacingly underneath the town which will explode spectacularly as soon as there is any kind of leak. This scenario was emphatically not considered in any risk assessment partly because such an explosion would be contrary to anything we know about the behavior of stored CO_2, and it is probably fair to say that therefore it is not included in risk assessments nor even considered as a known inadequacy of the models. Nevertheless, this scenario of something going horribly wrong somehow is a valid concern especially for those who do not possess the expert background knowledge to adequately judge it as so unlikely as to not even be worth including in the model.

This rough overview on the risk debates on CO_2 storage is meant to show that there are very much different perspectives we can take on the risk of CCS, and which ones we put most stock in depend on our background knowledge and ideologies. As in the red meat and cancer example above, different actors in the debate have emphasized different types of uncertainties in this case. Energy companies like Shell or BP who are developing CCS projects, as well as those politicians who are keen on promoting it take comfort from the fact that final risk assessments put the safety risks of the technology as very low, and in much of the industry communication literature on CCS, it is these figures that get mentioned rather than any more technical discussion surrounding how they were acquired. Literature from environmental groups on the other hand see the uncertainties in a different context by highlighting the potential of measurement errors in local evaluations, casting doubt on the modeling processes involved (for example, Greenpeace 2008).

Climate Change

This leads us to a final very brief example because the one thing that complicates debates about CCS is its relation to the mitigation of climate change. Man-made climate change presents a particular problem, not because there are by now any doubts left that it is happening, but because the forecasting of how bad it will be under various scenarios is a very imprecise business. Scientists who try to predict possible climate futures almost always start by admitting that they are working from one particular model, and that there are several that we know we could use instead that give different estimates, and we rarely have any good handle of working out how likely each model is to reflect reality. In fact, it is often acknowledged that we know so little and climate and weather patterns are so complex that there are always possible factors that we do not even know that we do not know about which can take us by complete surprise.

Uncertainties in climate change modeling are thus dominated by levels 4 and even 5: We have several models to choose from, but not much knowledge on how well they are doing their jobs, and even then we are well aware that our forecasting is hostage to completely unforeseen things as well. Social science research on climate modelers themselves has shown that there is a wide range of expert opinions on the best modeling process and that, moreover, experts themselves will have an unreliable estimation as to the possible shortcomings of their own models (Lahsen 2005). Even then, if they have an adequate estimation of the reliability of their models, experts find it hard to communicate them, especially when these estimations cannot be quantified (Hillerbrand 2009). Therefore, yet again, for the nonexpert observer of these debates, the most important uncertainty is not over which model is best, but over which expert to trust most, and because there is so much disagreement and unreliability of the experts' own assessments of their models, this uncertainty weighs more for the nonexpert than for the experts.

Further Research

By conceptualizing uncertainty along the various dimensions introduced above we can gain an appreciation of what the different disciplines find interesting about risk and, hopefully, find how they interconnect. The risk society literature, for example, as I outlined above is interested in different aspects of risk than the risk management literature. Considering the particular dimension that I think is most important in my map, the objects of uncertainty, gives us an idea of how and why different people estimate uncertainties differently even when presented with the same information and furthermore shows how different disciplines can themselves study risk from different perspectives.

This is therefore more useful for sociological research than the otherwise admirable combined system of van Asselt and Rotmans (2002) and Walker et al. (2003), whose scheme was designed primarily for use by experts in integrated assessments. Unlike Wynne's and Stirling's systems, which were designed for sociologists to understand and analyze different reactions to uncertainty, my map and classification tries to be more inclusive and meticulous in teasing apart different aspects of uncertainty, while hopefully still being simple enough to be useful for gaining an immediate and intuitive understanding of why and how opinions on risk so often differ. This can then become a useful tool when social scientists communicate the problems of the social contexts of risk to technical experts—this is after all a role performed very often by social scientists who have been funded by scientific research institutions and funding agencies "to look at the social side of things," but who often struggle to make the social insights relevant and intuitive to the technical experts they work with. This is a main reason why we (Riesch and Reiner 2010; Upham et al. 2011) use this framework in the study of risk opinions on energy infrastructure (CCS and biofuels, respectively), to so far very positive reactions from the technical communities. This approach therefore tries to marry the sociological usefulness of Wynne with the technical relevance of Walker and van Asselt and their colleagues.

Though the scheme presented here is meant to be illustrative rather than prescriptive, it shows some clear lessons for risk communication strategies. Since, as I have argued here, different people are worried about different aspects of the risks and, in particular, attach the uncertainty to different objects, risk communication strategies often fail to convey the information that people actually find important. While there is no silver bullet with which to persuade people who simply do not trust the experts, or who's understanding of technological risks gives a higher importance to unforeseen events, taking these different perspectives into account will ensure that the conversation at least does not disintegrate into different actors failing to understand each other. In designing a communication tool about the risks of CCS, for example, we may want to pay particular attention not just to the risks as calculated by the risk assessment, but also how it was arrived at, what the uncertainties with the parameters are, what was the choice of models available, and why was this particular one chosen, what possible inadequacies were not modeled and finally what are the plans for action should unforeseen consequences occur.

Future research will hopefully develop some of the other dimensions along the more detailed level as I have tried with the objects of uncertainty dimension (and as van Asselt and Rotmans have already done with the sources of uncertainty). This would then allow us to construct a more detailed table through which we can map, at a glance, where the different academic literatures on risk lie and intersect and which may help researchers in finding connections and future ideas for more integrated interdisciplinary research on risk.

Work is also underway to develop case studies which apply the objects classification to different risk scenarios. I have summarized here the application to CCS (Riesch and Reiner 2010); furthermore, we are applying it to the problem of indirect land use change for the biomass energy industry (Upham et al. 2011). Furthermore, detailed studies on more diverse risk situations will hopefully be able to tease out more of the potential but also limitations of our scheme.

References

Aven T, Renn O (2009) On risk defined as an event where the outcome is uncertain. J Risk Res 12(11):1–11

Bailer-Jones DM (2003) Scientists' thoughts on scientific models. Perspect Sci 10:275–301

Beck U (1992) Risk society: towards a new modernity. Sage, London

Campbell S, Currie G (2006) Against beck: in defence of risk analysis. Philos Soc Sci 36(2): 149–172

Carlyle T (1871) Oliver cromwell's letters and speeches: with elucidations. Scribner, Welford, New York. http://www.gasl.org/refbib/Carlyle__Cromwell.pdf. Accessed 1 Sep 2010

Douglas M (1992) Risk and blame. Routledge, London

Edwards A, Elwy G, Covey J, Matthews E, Pill R (2001) Presenting risk information—a review of the effects of "framing" and other manipulations on patient outcomes. J Health Commun 6:61–82

EuropeanUnion (2010) Implementation of directive 2009/31/EC on the geological storage of carbon dioxide. http://ec.europa.eu/clima/policies/lowcarbon/docs/GD1-CO2%20storage %20life%20cycle%20risk%20management-consultation.pdf. Accessed 31Dec 2010

Funtowicz SO, Ravetz JR (1990) Uncertainty and quality in science for public policy. Kluwer, Dordrecht

Funtowicz SO, Ravetz JR (1993) Science for the postnormal age. Futures 25:739–755

Giddens A (1999) Risk and responsibility. Mod Law Rev 62:1–10

Gigerenzer G (2002) Reckoning with risk. Penguin, London

Gillies D (2000) Philosophical theories of probability. Routledge, London

Greenpeace (2008) False hope: why carbon capture and storage won't save the climate. Greenpeace International, Amsterdam

Hacking I (1975) The emergence of probability. Cambridge University Press, Cambridge

Hillerbrand R (2009) Epistemic uncertainties in climate predictions. A challenge for practical decision making. Intergenerational Just Rev 9(3):94–99

Hoeting J, Madigan D, Raftery A, Volinsky CT (1999) Bayesian model averaging: a tutorial. Stat Sci 14:382–417

Hogg MA (2007) Social identity and the group context of trust: managing risk and building trust through belonging. In: Gutscher H, Siegrist M, Earle TC (eds) Trust in cooperative risk management. Earthscan, London, pp 51–72

Itaoka K, Saito A, Akai M (2004) Public acceptance of CO2 capture and storage technology : a survey of public opinion to explore influential factors. In: Rubin ES, Keith DW, Gilboy CF (eds) Proceedings of 7th international conference on greenhouse gas control technologies, vol 31, 1: Peer-reviewed papers and plenary presentations, IEA Greenhouse Gas Program, Cheltenham

Jasanoff S (2003) Technologies of humility: citizen participation in governing science. Minerva 41:223–244

Jaspers A (2009) Slapen met de ramen dicht. Natuur-Wetenschap & Techniek (NWT) 77(4): 24–33

Jaspers A (2010) The view from technological journalism. FENCO workshop CCS and public engagement, Amsterdam

Joffe H (1999) Risk and the other. Cambridge University Press, Cambridge

Keynes JM (1937) The general theory. Q J Econ 51:209–233

Knight FH (1971) Risk, uncertainty and profit (reprint of the 1921 edn). University of Chicago Press, Chicago

Lahsen M (2005) Seductive simulations? Uncertainty distribution around climate models. Soc Stud Sci 35(6):895–922

Mander S, Polson D, Roberts T, Curtis A (2010) Risk from CO2 storage in saline aquifers: a comparison of lay and expert perceptions of risk. GHGT10, Amsterdam

Moscovici S (2000) Social representations. Polity Press, Cambridge

Norton J, Brown J, Mysiak J (2006) To what extent, and how, might uncertainty be defined? Comments engendered by Defining uncertainty: a conceptual basis for uncertainty management in model-based decision support: Walker et al., Integrated Assessment 4: 1, 2003. Integrated Ass J 6(1):83–88

Ravetz JR (2006) Post-normal science and the complexity of transitions towards sustainability. Ecol Complex 3:275–284

Raza Y (2009) Uncertainty analysis of capacity estimates and leakage potential for geologic storage of carbon dioxide in saline aquifers. Masters Thesis, MIT, Cambridge

Renn O, Klinke A (2004) Systemic risks: a new challenge for risk management. EMBO Rep 5(S1):S41–S46

Riesch H, Spiegelhalter DJ (2011) Careless pork costs lives: risk stories from science to press release to media. Health Risk Soc 13(1):47–64

Riesch H, Reiner D (2010) Different levels of uncertainty in carbon capture and storage (submitted)

Roeser S (2009) The relation between cognition and affect in moral judgments about risk. In: Asveld L, Roeser S (eds) The ethics of technological risk. Earthscan, London, pp 182–201

Roeser S (2010) Intuitions, emotions an gut reactions in decisions about risks: towards a different interpretation of neuro ethics. J Risk Res 13(2):175–190

Rumsfeld D (2002) Defense.gov news transcript. http://www.defense.gov/transcripts/transcript.aspx?transcriptid=2636. Accessed 3 Dec 2010

Schneider SH, Turner BL, Garriga HM (1998) Imaginable surprise in global change science. J Risk Res 1(2):165–185

Shackley S, Wynne B (1996) Representing uncertainty in global climate change science and policy: boundary ordering devices and authority. Sci Technol HumVal 21(3):275–302

Slovic P (2000) The perception of risk. Earthscan, London

Slovic P, Finucane M, Peters E, MacGregor D (2004) Risk as analysis and risk as feelings: some thoughts about affect, reason, risk and rationality. Risk Anal 24(2):311–322

Sorensen R (2009) Epistemic paradoxes. In: Zalta E (ed) The Stanford encyclopedia of philosophy. http://plato.stanford.edu/archives/spr2009/entries/epistemicparadoxes/. Accessed 31 Dec 2010

Spiegelhalter DJ (2010) Quantifying uncertainty. In: Paper presented at handling uncertainty in science, Royal Society, London, 22–23 Mar 2010

Spiegelhalter DJ, Pearson M (2008) 2845 ways to spin the risk. http://understandinguncertainty.org/node/233. Accessed 31 Dec 2010

Spiegelhalter DJ, Riesch H (2011) Don't know, can't know: embracing deeper uncertainties when analysing risks. Philos T Roy Soc A 396(1956):4730–4750. doi:10.1098/rsta.2011.0163

Stern N (2007) The economics of climate change. Cambridge University Press, Cambridge

Stirling A (2007) Risk, precaution and science: towards a more constructive policy debate. EMBO Rep 8(4):309–315

Sunstein CR (2005) Laws of fear. Cambridge University Press, Cambridge

Tajfel H (1981) Human groups and social categories: studies in social psychology. Cambridge University Press, Cambridge

Taleb N (2007) The black swan: the impact of the highly improbable. Penguin, London

Upham P, Riesch H, Tomei J, Thornley P (2011) The sustainability of woody biomass supply for UK bioenergy: a post-normal approach to environmental risk and uncertainty. Environ Sci Policy 14(5):510–518

van Asselt MBA, Rotmans J (2002) Uncertainty in integrated assessment modelling: from positivism to pluralism. Clim Change 54:75–105

Walker WE, Harremoes P, Rotmans J, van der Sluijs JP, van Asselt MBA, Janssen P, Krayer von Krauss MP (2003) Defining uncertainty: a conceptual basis for uncertainty management in model-based decision support. Integrat Ass 4(1):5–17

Washer P (2004) Representations of SARS in the British newspapers. Soc Sci Med 59:2561–2571

WCRF (2007a) Food, nutrition, physical activity, and the prevention of cancer: a global perspective. WCRF, Washington

WCRF (2007b) Landmark report: excess body fat causes cancer. http://www.wcrf-uk.org/press_media/releases/31102007.lasso. Accessed 13 Aug 2008

WCRF (2007c) Media quotes. http://www.wcrf-uk.org/press_media/quotes.lasso. Accessed 13 Aug 2008

Wynne B (1992) Uncertainty and environmental learning: reconceiving science and policy in the preventive paradigm. Global Environ Chang 2(2): 111–127

Chapter 3
The Role of Feelings in Perceived Risk

Melissa L. Finucane

Abstract This chapter provides an overview of key conceptualizations of and evidence for the role of feelings in perceived risk. Influence from feelings in judgment and decision making was first recognized nearly three decades ago. More recent work has developed models that generalize the mechanisms by which feelings operate. Feelings may play multiple roles in judgment and decision processes, including providing information, enabling rapid information processing, directing attention to relevant aspects of the problem, facilitating abstract thought and communication, and helping people to determine social meaning and to act morally. Feelings may be anticipated or experienced immediately and either integral (attached) to mental representations of the decision problem or incidental (unrelated), arising from moods or metacognitive processes. A rich repertoire of psychological concepts related to risk, such as appraisal and memory, can be used to help explain the mechanisms by which affect and analysis might combine in judgment and decision making. Phenomena such as psychophysical numbing, probability neglect, scope insensitivity, and the misattribution of incidental affect all provide empirical support, albeit fragmented, for the important influence of feelings. Future research needs to utilize multiple dependent variables and methodological approaches to provide convergent evidence for and development of more sophisticated descriptive and predictive models. An additional direction for future research is to develop tools that help risk communicators and risk mangers to address complex, multidimensional risk problems.

M. L. Finucane (✉)
East–West Center, Honolulu, HI, USA
e-mail: Melissa.Finucane@EastWestCenter.org

S. Roeser et al. (eds.), *Essentials of Risk Theory*, SpringerBriefs in Philosophy,
DOI: 10.1007/978-94-007-5455-3_3, © The Author(s) 2013

Introduction

The study of the role of feelings in risk judgments began with a focus on regret and disappointment theories within an economic framework (Bell 1982; Loomes and Sugden 1982) and experimental manipulations of mood (Johnson and Tversky 1983; Isen and Geva 1987). Nearly three decades later, researchers have amassed considerable evidence recognizing the importance of feelings in shaping risk perceptions. Numerous approaches have been used to capture and explicate feelings-based processes in a wide variety of domains. Research has moved on from establishing that feelings play a role, to developing models that generalize the mechanisms by which risk perceptions are influenced (Pham 2007; Slovic 2010). This chapter provides an overview of key conceptualizations of and evidence for the role of feelings in risk judgments. An intentionally wide-ranging use of the term "feelings" is employed to include studies of affect, emotion, and mood, reflecting the diverse theories and methods that comprise this field of research.

Conceptualizations

Dual-Process Theories: Recognizing Reliance on Feelings

Neoclassical economics asserts that individuals, over time and in aggregate, process risk information only in a way that maximizes expected utility (von Neumann and Morgenstern 1947). From this perspective, judgments are based on a utilitarian balancing of risks and benefits and feelings are only a byproduct of the cognitive process. That is, emotions such as fear, dread, anger, hope, or relief are experienced *after* the risk–benefit calculation is complete.

More recently, dual-process theories have conceptualized perceptions of and responses to risk as typically reflecting two, interacting, information-processing systems (Damasio 1994; Epstein 1994; Sloman 1996; Kahneman 2003; Bechara and Damasio 2005). The "analytic" system reflects the slow, deliberative analysis we apply to assessing risk and making decisions about how to manage hazards. The "experiential system" reflects fast, intuitive, affective reactions to danger. "Affective reactions" refer to a person's positive or negative feelings about specific objects, ideas, images, or other target stimuli. Feelings may also reflect emotions (intense, short-lived states of arousal accompanied by expressive behaviors, specific action tendencies, and conscious experiences, usually with a specific cause, Forgas 1992) and moods (feelings with low intensity, lasting a few minutes or several weeks, often without specific cause, Isen 1997). From the dual-process perspective, feelings that arise from or amidst the experiential mode of thinking are influential during judgment and decision-making processes (Schwarz and Clore 1988).

Reliance on feelings in the process of evaluating risk has been termed "the affect heuristic" (Finucane et al. 2000a). Feelings provide potentially useful inputs to judgments and decisions, especially when knowledge about the events being considered is not easily remembered or expressed (Damasio 1994). Many theorists have also given feelings a direct and primary role in motivating and regulating behavior (Mowrer 1960; Zajonc 1980; Damasio 1994; Isen 1997; Kahneman 2003; Pham 2007). Positive feelings act like a beacon of incentive, motivating people to act to reproduce those feelings, whereas negative feelings motivate actions to avoid those feelings.

Since recognizing the importance of feelings, scholars have attempted to clarify the nature and timing of their influence on risk perceptions. Distinguishing the impact of specific emotional states is of concern because the desirability of the impact may be a function of the intensity, valence, and appraisal content of the emotion. For instance, Lerner and Keltner (2001) have shown that fearful people express pessimistic risk estimates and risk-averse choices, whereas angry people express optimistic risk estimates and risk-seeking choices. Similarly, the importance of anticipated regret and disappointment has been demonstrated by Zeelenberg et al. (2000) and Connolly and Butler (2006). The timing of feelings is also critical. In analyses of the time course of decisions, Loewenstein and Lerner (2003) distinguish between anticipated emotions (beliefs about one's future emotional states that might ensue after particular outcomes) and immediate emotions (experienced when making a decision, thereby exerting an influence on the choice process). (For similar distinctions see Kahneman 2000.) Loewenstein and Lerner further identified two types of immediate emotions, namely, integral emotions (caused by the decision problem itself, such as feelings about a target stimulus or available options) and incidental emotions (caused by factors unrelated to the decision problem at hand, such as mood or cognitive fluency); see also Bodenhausen (1993) and Pham (2007). Empirical demonstrations of the influence of integral and incidental feelings on a wide variety of judgments and decisions are reviewed below.

In sum, early models of judgment and decision making emphasized cognitive aspects of information processing and viewed feelings only as a byproduct of the cognitive process. More recent models, however, give feelings a direct and primary role in motivating and regulating behavior in response to risk. Feelings may be anticipated or immediate and either integral to the decision problem or incidental, arising from moods or metacognitive processes. Identifying the role of feelings and how they interact with cognitive processes is the current focus of scientific inquiry for many researchers.

Functional Frameworks: Identifying the Roles of Feelings

In recent work, Peters (2006) proposed a framework to capture four roles that feelings play in judgment and decision processes. The first role is to provide information about the target being evaluated. Based on prior experiences relevant to choice options (integral affect) or the result of less relevant and ephemeral states (incidental affect), feelings act as information to guide the judgment or decision process (Slovic et al. 2002). The second role is as a spotlight. The extent or type of feelings (e.g., weak vs. strong or anger vs. fear) focuses the decision maker's attention on certain kinds of information, making it more accessible for further processing. Third, feelings may operate as a motivator of information processing and behavior, influencing approach-avoidance tendencies (Frijda et al. 1989; Zeelenberg et al. 2008). Incidental mood states may also motivate people to act in a way that maintains a positive mood (Isen 2000). A fourth role is to serve as a common currency in judgments and decisions, allowing people to compare disparate events and complex arguments on a common underlying dimension (Cabanac 1992). Integrating good and bad feelings is easier than trying to integrate multiple incommensurate values and disparate logical reasons. A similar functional framework has been proposed by Pfister and Böhm (2008), who emphasize the role of feelings in providing information, directing attention to relevant aspects of the problem, and enabling rapid information processing.

An additional function of feelings, according to Pfister and Böhm (2008), is to generate commitment to implementing decisions, thus helping people to act morally, even against their short-term self-interest. Roeser (2006, 2009, 2010) also highlights the importance of emotions in providing moral knowledge about risks and that emotions are needed to correct immoral emotions. Kahan (2008) describes emotion as providing "a perceptive faculty uniquely suited to discerning what stance toward risk best coheres with a person's values." In his cultural evaluator theory, Kahan regards emotion as entering into risk judgments as a way of helping people to evaluate the social meaning of a particular activity against a background of cultural norms and to express the values that define their identities. When people draw on their feelings to judge risk, they form an attitude about what it would mean for their cultural worldviews for society to agree that the risk is dangerous and worthy of regulation. Kahan distinguishes the role of feelings not as a heuristic but as unique in enabling a person to identify a stance that is "expressly rational" for someone with commitment to particular worldviews. Consistent with his theory, Kahan et al. (2007) found that the impact of affect relative to other influences (such as gender, race, or ideology) was significantly larger among people who knew a modest or substantial amount about nanotechnology. This contrasts with the heuristic perspective in which affect is expected to play a larger role when someone lacks sufficient information to form a coherent judgment.

Finally, feelings may help to facilitate abstract thought and communication (Finucane and Holup 2006). Feelings help people to think abstractly because they link abstract concepts (e.g., good, bad) to the physical or sensory world. Without such links, judgments are slower and less accurate. One subtle demonstration of the link between affect and analytic thought is research showing that positive words are evaluated faster and more accurately when presented in white font, whereas negative words are evaluated faster and more accurately when presented in black font, despite the brightness manipulation being orthogonal to the valence of the words (Meier et al. 2004). Similarly, Meier and Robinson (2004) showed that people assign "goodness" to objects high in visual space and "badness" to objects low in visual space. Linking abstract concepts to physical or sensory experiences helps the analytic system to interpret the meaning of stimuli so that they can be incorporated in cognitive calculus.

In sum, several roles of feelings in judgments and decisions about risk have been identified. Additional roles may be articulated as diverse disciplines apply their perspectives. Which role dominates in any particular judgment or decision is likely to be a function of multiple factors (e.g., task demands, time pressure, preferred decision style, social norms).

Clarifying the Relationship Between Feelings-Based and Cognitive Processes

Despite recognition that feelings-based and cognitive processes represent inter-dependent systems in decision making (Damasio 1994; Epstein 1994; Sloman 1996; Kahneman 2003; Bechara and Damasio 2005), theory and research to date have struggled to convey the exact nature of the relationship. The cognitive origins of behavioral decision theory may have encouraged people to assume that the domain of feelings is qualitatively different and functionally separate from the domain of cognition. Such distinction is reflected in the dichotomies often portrayed in this field, such as irrational emotions disturbing rational cognitions, intuitive feelings dominating deliberate thinking, and hot affect overwhelming cold logic (Pfister and Böhm 2008). However, overlapping commonalities in the systems have been noted also. For instance, the processing of experiences may be involved in both affective and analytic approaches. Johnson and Lakoff (2002; Lakoff and Johnson 1999) point out that even our most abstract thinking (mathematics, for example) is based on our "embodied" experiences. They describe how the locus of experience, meaning, and thought is the ongoing series of physical interactions with our changing environment. Our embodied acts and experiences are an important part of our conceptual system and in making sense of what we experience.

Clarifying the mechanisms by which feelings and cognitions are related and integrated in human judgment and decision making is a critical next step in understanding perceived risk. Finucane and Holup (2006) recommend expanding and linking the risk-as-analysis and risk-as-feelings approaches by adopting a "risk-as-value" model. This model emphasizes that responses to risk result from a combination of analysis and affect that motivates individuals and groups to achieve a particular way of life. Derived from dual-process theories, the risk-as-value model implies that differences in perceived risk may arise from differences in the analytic or affective evaluation of a risk or the way these evaluations are combined. As research moves from simply describing variance to predicting it, having multiple potential loci for such variation with different substantive interpretations will be useful. The risk-as-value model does not posit a specific rule for combining affective and analytic evaluations, although traditional information integration rules (adding, averaging, multiplying) may be applicable in some contexts. When the implications of both affective and analytic evaluations are congruent, the processes may be more likely to combine additively. However, incongruence may result in greater emphasis on analytic or affective processing, depending on an array of task, decision-maker, or context variables (e.g., analysis may be increased if it is viewed as more reliable, but may be attenuated under time pressure).

The relationship between affective and analytic processes may be more fully explained by drawing on the rich repertoire of empirically tested concepts related to the psychology of risk, such as appraisal and memory. Lerner and Keltner's (2000) appraisal-tendency theory suggests that emotions arise from but also elicit specific cognitive appraisals. For instance, fear arises from and evokes appraisals of uncertainty and situational control, whereas anger is associated with appraisals of certainty and individual control (Lerner and Keltner 2001). Lerner et al. (2003) showed that anger evokes more optimistic beliefs about risks such as terrorism, whereas fear evokes greater pessimism about risks. Weber and Johnson's (2006) preferences-as-memory framework highlights how risk judgments are made by retrieving relevant (cognitive and affective) knowledge from memory. Framing normatively equivalent information positively or negatively (e.g., 90 % lives saved vs. 10 % lives lost) influences preferences because the different descriptions prime different representations in memory (predominantly positively or negatively valenced). Also drawing on modern concepts of memory representation, retrieval, and processing, Reyna and colleagues (Reyna and Brainerd 1995; Reyna et al. 2003) have proposed a dual-process model called fuzzy-trace theory (FTT). FTT posits that people form two kinds of mental representations. The first, verbatim representations are detailed and quantitative. The second, gist representations, provide only a fuzzy trace of experience in memory. People tend to rely primarily on gist, which captures the meaning of experience, including the emotional meaning. FTT differs from other dual-process models by placing intuition at the highest level of development, viewing fuzzy intuitive processes as more advanced than precise analytic processes (Reyna 2004).

In sum, the mechanisms by which feelings and cognitions are combined in judgment and decision making need to be clarified. Studies from a wide range of disciplines, including cognitive and social psychology, emotion and motivation, economics, decision research, and neuroscience, need to be integrated to develop models that explicitly specify possible causal constructs or variables that influence reactions to risk, allow for individual and group differences in these variables or in the relationships between them, and generalize across risk domains and contexts. Such model-based research can broaden our understanding of risk perceptions specifically and of basic psychological phenomena more generally.

Empirical Support

This section briefly reviews empirical support for the role of feelings in risk judgments and decisions. Although the empirical literature seems fragmented and sometimes inconsistent, evidence for the influence of emotion, affect, and mood is compelling.

Integral Feelings as a Proxy for Value

Early evidence of the role of feelings in risk perceptions came from studies showing that "dread" was the major driver of public acceptance of risk in a wide range of contexts, including environmental hazards such as pesticides, coal burning (pollution), and radiation exposure from nuclear power plants (Fischhoff et al. 1978). This observation led to many studies looking at how risk judgments are influenced by feelings that are integral (attached) to mental representations of hazardous activities, technologies, or events (Loewenstein et al. 2001; Slovic et al. 2002). In the first paper published on the affect heuristic, Finucane et al. (2000a) demonstrated that providing information about benefit (e.g., of nuclear power) changed perceptions of risk and vice versa. They also showed that whereas risk and benefit (e.g., of natural gas, chemical fertilizers) tend to be positively correlated across hazards in the world, they are negatively correlated in people's judgments. Moreover, this inverse relationship between perceived risks and benefits increased greatly under time pressure, a situation in which opportunity for analytic deliberation was reduced. Although subconscious cognitive processes cannot be ruled out entirely, these results support the notion that in the process of judging risk, people may rely on feelings as a source of information about whether or not they are at risk and how they should respond.

Underpinning processes such as the affect heuristic are images, to which positive or negative feelings become attached through learning and experience. Images include perceptual representations (pictures, sounds, smells) and symbolic representations (words, numbers, symbols) (Damasio 1999). In an influential series of studies using the Iowa Gambling Task, Damasio, Bechara, and colleagues (Bechara et al. 1994; Damasio 1994; Bechara et al. 1997) proposed that in normal individuals, emotional responses evoked by objects are stored with memory representations (images) as somatic markers of these objects' value (for challenges to the original interpretation, however, see Maia and McClelland 2004; Fellows and Farah 2005). Other research suggests that more vivid, emotionally gripping images of harm are more salient than emotionally sterile images, making those risks more likely to be noticed, recalled, and responded to (Hendrickx et al. 1989; Sunstein 2007). One explanation for this vividness effect may be that initial affective responses to an object seem to trigger a confirmatory search for information that supports the initial feelings (Pham et al. 2001; Yeung and Wyer 2004), possibly increasing the subjective coherence of judgments based on affect (Pham 2004). Another explanation may relate to the inherently strong drive properties of integral feelings, which motivate behavior and redirect action if necessary (Frijda 1988).

A simple method for studying the relationship between affect, imagery, and perceived risk is called affective image analysis, a structured form of word association and content analysis (Slovic et al. 1991; Benthin et al. 1995; Finucane et al. 2000b; Jenkins-Smith 2001; Satterfield et al. 2001). This method allows researchers to examine the distribution of different (sometimes conflicting) meanings of risk across people and to identify and explain those images that carry a strongly positive or negative emotional charge. For instance, Finucane et al. (2000a) asked study participants to free associate to the phrase "blood transfusions." Associations included "HIV/AIDS," "hemophilia," "gift giving," and "life saving." Participants were then asked to rate each of their associations on a scale from bad (−3) to good (+3); these ratings were correlated with a number of other measures, such as acceptability of having a transfusion and sensitivity to stigmatization in other risk settings. Affective image analysis was also used in a US national survey by Leiserowitz (2006), who found that holistic negative affect and image affect were significant predictors of global warming risk perceptions, explaining 32 % of the variance. Holistic negative affect was also predictive of support for national policies to address global warming, but less predictive than worldviews and values. A content analysis of affective imagery associated with "global warming" revealed that the phrase evoked negative connotations for almost all respondents, but that the most dominant images referred to impacts that were psychological or geographically distant, generic increases in temperature, or a different environmental problem.

In sum, integral affective responses are feelings elicited by real, perceived, or imagined images of the object of judgment or decision. These feelings are predictive of a variety of behavioral responses to risk. Evaluation and choice processes are more likely to be influenced by vivid, emotionally gripping images than pallid representations, possibly because strong feelings trigger a confirmatory information search or strong drive states.

Psychophysical Numbing

Considerable evidence suggests that affective responses follow the same psychophysical function that characterizes our sensitivity to a range of perceptual stimuli (e.g., brightness, loudness). In short, people's ability to detect changes in a physical stimulus decreases as the magnitude of the stimulus increases. Known as Weber's law, the just-noticeable change in a stimulus is a function of a fixed percentage of the stimulus. That is, to notice a change, only a small amount needs to be added to a small stimulus, but a large amount needs to be added to a large stimulus (Stevens 1975). Our cognitive and perceptual systems are designed to detect small rather than large changes in our environment. Fetherstonhaugh et al. (1997) demonstrated this same phenomenon of psychophysical numbing (i.e., diminished sensitivity) in the realm of feelings by evaluating people's willingness to fund alternative life-saving medical treatments. Study participants were asked to indicate the number of lives a hypothetical medical research institute would have to save to merit a $10 million grant. Nearly two-thirds of participants raised their minimum benefit requirements to warrant funding when the at-risk population was larger. A median value of 9,000 lives needed to be saved when 15,000 were at risk, compared with a median of 100,000 lives when 290,000 were at risk. In other words, 9,000 in the smaller population seemed more valuable than saving ten times as many lives in the larger population. Psychophysical numbing or proportional reasoning effects have been demonstrated also in other studies (Baron 1997; Friedrich et al. 1999).

In striving to explain when feelings are most influential in judgments about saving human lives, several researchers have explored the "identifiable victim effect" (Jenni and Loewenstein 1997; Kogut and Ritov 2005; Small and Loewenstein 2005). For instance, Small et al. (2007) asked participants to indicate how much they would donate to a charity after being shown either statistical information about the problems of starvation in Africa ("statistical victims") or a photograph of a little girl in Africa and a brief description of the starvation challenges she faces ("identifiable victim"). Results showed that the mean donation ($2.83) for the identifiable victim was more than twice the mean donation ($1.17) for the statistical victim, as might be expected given the affectively engaging nature of the photograph of the identifiable victim. Most interestingly, however, when participants were shown both statistical and identifiable information simultaneously, the mean donation was $1.43. When jointly evaluating statistics and an individual victim, the reason for donating seems to become less compelling, possibly because the statistics diminish reliance on affective reactions during decision making. Small et al. also measured feelings of sympathy toward the cause (the identified or statistical victims). The correlation between these feelings and donations was strongest when people faced the identifiable victim.

In a follow-up study by Small et al. (2007), participants were either primed to feel ("Describe your feelings when you hear the word 'baby'") or to deliberate ("If an object travels at five feet per minute, how many feet will it travel in 360 s?"). Relative to the feelings prime, priming deliberative thinking reduced donations to the identifiable victim. There was no discernible difference of the two primes on donations to statistical victims, as would be expected because of the difficulty in generating feelings for such victims. Similarly, Hsee and Rottenstreich (2004) demonstrated that priming analytic evaluation led to more scope sensitivity and affective evaluation led to more scope insensitivity when participants were asked how much they would be willing to donate to help save endangered pandas. In their study, the number of pandas was represented in an affect-poor manner (i.e., as large dots) or an affect-rich manner (i.e., with a cute picture). The dots were related to a fair degree of scope sensitivity (mean donations were greater for four pandas than one), whereas pictures were related to scope insensitivity (mean donations for four versus one panda were almost identical). This scope insensitivity violates logical rationality, suggesting that inherent biases in the affective system can lead to faulty judgments and decisions.

In sum, the affective system seems designed to be most sensitive to small changes at the cost of making us less able to respond appropriately to larger changes further away from zero. Consequently, we may fail to respond logically to humanitarian and environmental crises.

Nonintuitive Consequences

Integral affect may lead decision makers astray in several other ways. One example is the phenomenon of "probability neglect"—the failure of people to adjust their decisions about the acceptability of risks to changes in information about their probability. Loewenstein et al. (2001) observed that responses to uncertain situations appear to have an all-or-none characteristic, sensitive to the possibility of strong negative or positive consequences and insensitive to their probability. That is, strong feelings tend to focus people on outcomes rather than probabilities. Rottenstreich and Hsee (2001) demonstrated that while people were willing to pay more to avoid a high than a low probability of losing $20, they were not willing to pay more to avoid a high than a low probability of receiving an electric shock (a prospect rich in negative affect). Another example comes from Denes-Raj and Epstein's (1994) jellybeans experiment. When given a chance to draw a winning red bean either from a small bowl containing a single red bean and nine white beans (10 % chance of winning) or from a larger bowl containing nine red beans and 91 white beans (9 % chance of winning), people tend to choose to draw from the larger bowl, even though the probability of winning is greater with the small bowl. The more abstract notion of probability (the distribution of beans in a random draw process) is less influential than the affective response people have to the concrete representations of objects (seeing multiple red beans). One interpretation of these results is that integral affect provides a largely categorical approach to assessing value. That is, objects are categorized in terms of their significance for well-being, regardless of their probability or magnitude.

The emerging field of neuroeconomics provides convergent evidence for the nonintuitive consequences of integral affect (Trepel et al. 2005). Using methods such as functional magnetic resonance imaging, researchers have examined brain activity in areas known to process affective information. For instance, examining the neurobiological substrates of dread, Berns et al. (2006) showed that when people are confronted with the prospect of an impending electric shock, regions of the pain matrix (a cluster of brain regions activated during a pain experience) are activated. This finding suggests that people not only dislike experiencing unpleasant outcomes, they also dislike waiting for them. Contrary to tenets of economic theory, people seem to derive pain (and pleasure) directly from information, rather than from any material outcome that the information might lead to. Anticipating future outcomes in this way can have a major impact on intertemporal choices (decisions that involve costs and benefits that extend over time). While an economic account of intertemporal choice predicts that people generally want to expedite pleasant outcomes and delay unpleasant ones (Loewenstein 1987), an affective account predicts that people may prefer to defer pleasant outcomes when waiting is pleasant or to expedite unpleasant outcomes when waiting is frustrating or produces dread.

Another nonintuitive feature of feelings-based judgments is that they tend to be more relativistic or reference-dependent than are reason-based judgments. That is, affective responses are often not based on the object or outcome in isolation, but in relation to other objects or outcomes (Mellers 2000). Winning $10 in a gamble will elicit greater pleasure if the alternative outcome is losing $5 rather than only $1. This finding is also consistent with work on the evaluability principle. Hsee (1996) asked people to assume they were music majors looking for a used music dictionary. Participants were shown two dictionaries and asked how much they would be willing to pay for each. Dictionary A had 10,000 entries and was like new, whereas dictionary B had 20,000 entries but also had a torn cover. In a joint-evaluation condition, willingness to pay was higher for B, presumably because of its greater number of entries. However, when one group of participants evaluated only A and another group evaluated only B, the mean willingness to pay was much higher for A, presumably because without a direct comparison, the number of entries was hard to evaluate whereas the defects attribute was easy to translate into a precise good/bad response. Wilson and Arvai (2006) have extended this work to show that in some contexts, enhanced evaluability may not be sufficient to deflect attention away from the affective impressions of the choice pair and toward other decision-relevant risk information, a behavior they call affect-based value neglect.

In sum, strong feelings can lead people to ignore probabilities and magnitudes, possibly because in some situations integral affect can provide only a categorical and reference-dependent approach to valuation. Risk theory and practice will benefit from further explorations of the conditions under which feelings influence attention to and use of different types of information.

Misattribution of Incidental Feelings

In addition to studies focusing on integral feelings, a large number of studies have shown that affective states unrelated to the judgment target (incidental feelings) may influence judgments and decisions (Schwarz and Clore 1983; Isen 1997). An early study by Johnson and Tversky (1983) demonstrated that experimental manipulation of mood (induced by a brief newspaper report on a tragic event such as a tornado or flood) produced a pervasive increase in frequency estimates for many undesirable events, regardless of the similarity between the report and the estimated risk. More recently, Västfjäll et al. (2008) showed that eliciting negative affect in people by asking them to think about a recent major natural disaster (the 2004 tsunami) influenced judgments when the affect was considered relevant (e.g., the perceived risk of traveling to areas affected by the disaster), but also when it was not relevant (e.g., developing gum problems).

In a classic study, Schwarz and Clore (1983) demonstrated that people reported higher levels of life satisfaction when they were in a good mood as the result of being surveyed on a sunny day than people who were in a bad mood as a result of being surveyed on a rainy day. People incorrectly attributed their incidental moods as a reflection of how they felt about their personal lives. In general, the misattribution of incidental feelings to attentional objects tends to distort beliefs in an assimilative fashion. However, research suggests that the influence of incidental affect is neither stable nor unchangeable. Rather, it is a constructive process in which the decision maker needs to determine whether their feelings are a reliable and relevant source of information (Pham et al. 2001; Clore and Huntsinger 2007). For instance, Schwartz and Clore were able to reduce the influence of mood on participants' judgments of well-being with a simple reminder about the cause of their moods (e.g., sunny vs. cloudy weather), presumably triggering people to question the diagnostic value of the affective reaction for the judgment. Importantly, the manipulation changed the diagnostic value of the affective reaction, not the affective reaction itself (Schwarz 2004). Västfjäll et al. (2008) also demonstrated that manipulating the ease with which examples of disasters come to mind can influence risk estimates. Asking participants who had been reminded of the 2004 tsunami to list few (vs. many) natural disasters led to more pessimistic outlooks (measured via an index averaging judgments of the likelihood of positive and negative events), presumably because listing many natural disasters rendered incidental affect relatively less diagnostic for judgments.

Incidental affective states have been shown also to influence the nature of information processing most likely to occur. Negative mood states generally promote a more analytic form of information processing, whereas positive moods generally promote a less systematic, explorative form of processing. From an evolutionary perspective, negative moods may highlight a discrepancy between a current and desired state, signaling a need to analyze the environment carefully (Higgins 1987). Positive moods, on the other hand, may encourage variety seeking in order to build future resources (Fredrickson 1998). Empirical findings are not entirely consistent, however. Both positive and negative moods have been related to increased and decreased systematic processing (Isen and Geva 1987; Mackie and Worth 1989; Schwarz 1990; Baron et al. 1992; Gleicher and Petty 1992; Wegener and Petty 1994; Isen 1997).

In sum, incidental feelings may influence risk judgments and decisions. The diagnostic value of the feelings depends on the context. Fortunately, people can be primed to examine the diagnostic value of their feelings. Incidental feelings may also influence the extent to which individuals engage in systematic processing, although the exact nature of this relationship remains unclear.

Generalizations

Several generalizations can be made about the role of feelings in risk judgments. First, feelings in the form of emotions, affect, or mood can have a large impact on how risk information is processed and responded to. The multiple ways in which feelings influence risk judgments and decisions likely relate to several functions of feelings: providing information, focusing attention, motivating behavior, enabling rapid information processing, generating commitment to outcomes to help people act morally, and facilitating abstract thought and communication. Other functions may be identified with more in-depth explorations from diverse disciplinary perspectives on the relationship between feelings and perceived risk.

Feelings that are integral to objects are often interpreted as signals of the value of those objects, motivating people to approach or avoid accordingly. Assessments of value based on integral affect differ from cognitive assessments in that the feelings tend to be more categorical, reference dependent, and sensitive to vivid imagery. Consequently, judgments based on integral feelings may be insensitive to scale (probability or magnitude) and myopic, emphasizing immediate hedonic consequences (positive or negative) over future consequences. The influence of specific characteristics of feelings (e.g., valence, intensity) on judgment processes needs further investigation.

Milder incidental feelings that are unrelated to the judgment target are also influential in judgment processes. In seeking information to inform their judgments, people tend to use whatever is available to them at the time and sometimes misattribute their mood states or metacognitive experiences as a reaction to the target. A variety of interventions can help people discern the diagnostic value of feelings.

Further Research

Since empirical studies are designed in a specific theoretical and methodological context, no single study can fully answer the complex question of how feelings affect risk perceptions. However, to address the fragmented and sometimes inconsistent findings reported to date, future research needs to work to provide converging evidence for the role of feelings in judgment and decision processes. Converging evidence will be obtained by looking at multiple dependent variables and by using multiple methodological approaches to test alternative explanations of results (Weber and Hsee 1999). Though methods and measures for studying affect may be unfamiliar to many risk researchers, a wealth of tools exist in diverse disciplines studying the form and function of feelings. An interdisciplinary effort including physiological, neurological, psychological, sociological, and other approaches can be used to examine the interplay of affective and analytic processes in risk judgments, to yield the fullest understanding of risk reactions.

Future research also needs to explore new (e.g., qualitative) understandings of how affective and analytic processes (and their interactions) are best represented. A growing body of ethicists and social scientists have criticized purely quantitative approaches as ill-equipped to reflect public conceptualizations of the complex, multidimensional, and often nonmonetary qualities of risks being faced (Stern and Dietz 1994; Prior 1998; Satterfield and Slovic 2004; Finucane and Satterfield 2005; Roeser 2010). Likewise, the seemingly categorical, reference-dependent nature of the affective system may require new approaches to fully explicate nonintuitive consequences of feelings on risk judgments.

Another direction for future research is to evaluate the ecological validity of feelings. Adopting a Brunswikian (Brunswik 1952) approach, Pham (2004) suggests examining (a) the correlation between integral feelings elicited by objects and these objects' true criterion value (the ecological validity of feelings), and (b) the correlation between other available proxies of value and the object's criterion value (the ecological validity of alternative bases of evaluation). The ecological validity of incidental feelings could be examined in a similar fashion.

Finally, in a more practical realm, future research needs to help risk communicators and risk mangers to determine the most effective tools for presenting and processing risk information. For instance, research will help to make risk estimates more accurate and risk mitigation behaviors more timely if it informs us of how to make abstract probabilities meaningful, reduce the gap between anticipated and experienced affect, facilitate the integration of non-commensurate metrics, or engage ethical assessments. Practitioners from diverse fields such as health care services, food safety, terrorism prevention, environmental resource management, and disaster preparedness would benefit from a systematic translation of the rich body of research into practice. Tools that account for the role of feelings in a way that facilitates efficient yet sound decision making will enhance our ability to successfully regulate risks.

References

Baron J (1997) Confusion of relative and absolute risk in valuation. J Risk Uncertainty 14(3):301–309

Baron RS, Inman MB et al (1992) Emotion and superficial social processing. Motiv Emotion 16:323–345

Bechara A, Damasio A (2005) The somatic marker hypothesis: a neural theory of economic decision. Games Econ Behav 52:336–372

Bechara A, Damasio AR et al (1994) Insensitivity to future consequences following damage to human prefrontal cortex. Cognition 50:7–15

Bechara A, Damasio H et al (1997) Decision advantageously before knowing the advantageous strategy. Science 275:1293–1294

Bell DE (1982) Regret in decision making under uncertainty. Oper Res 30:961–981

Benthin A, Slovic P et al (1995) Adolescent health–threatening and health-enhancing behaviors: a study of word association and imagery. J Adolesc Health 17:143–152

Berns GS, Chappelow J et al (2006) Neurobiological substrates of dread. Science 312:754

Bodenhausen GV (1993) Emotions, arousal, and stereo-type-based discrimination: a heuristic model of affect and stereotyping, affect, cognition, and stereotyping. Academic, San Diego, pp 13–37

Brunswik E (1952) The conceptual framework of psychology. University of Chicago Press, Chicago

Cabanac M (1992) Pleasure: the common currency. J Theor Biol 155:173–200

Clore GL, Huntsinger JR (2007) How emotions information judgment and regulate thought. Trends Cogn Sci 11:393–399

Connolly T, Butler D (2006) Regret in economic and psychological theories of choice. J Behav Decis Mak 19:139–154

Damasio AR (1994) Descartes' error: emotion, reason, and the human brain. Avon, New York

Damasio A (1999) The feeling of what happens. Harcourt, Inc., New York

Denes-Raj V, Epstein S (1994) Conflict between intuitive and rational processing: when people behave against their better judgment. J Pers Soc Psychol 66:819–829

Epstein S (1994) Integration of the cognitive and the psychodynamic unconscious. Am Psychol 49:709–724

Fellows LK, Farah MJ (2005) Different underlying impairments in decision making following ventromedial and dorsolateral frontal lobe damage in humans. Cereb Cortex 15:58–63

Fetherstonhaugh D, Slovic P et al (1997) Insensitivity to the value of human life: a study of psychophysical numbing. J Risk Uncertainty 14(3):282–300

Finucane M, Holup J (2006) Risk as value: combining affect and analysis in risk judgments. J Risk Res 9(2):141–164

Finucane ML, Satterfield T (2005) Risk as narrative values: a theoretical framework for facilitating the biotechnology debate. Int J Biotechnol 7(1–3):128–146

Finucane ML, Alhakami A et al (2000a) The affect heuristic in judgments of risks and benefits. J Behav Decis Mak 13:1–17

Finucane ML, Slovic P et al (2000b) Public perception of the risk of blood transfusion. Transfusion 40:1017–1022

Fischhoff B, Slovic P et al (1978) How safe is safe enough? A psychometric study of attitudes toward technological risks and benefits. Policy Sci 9:127–152

Forgas JP (1992) Affect in social judgments and decisions: a multiprocess model. In: Zanna M (ed) Advances in experimental social psychology. Academic, San Diego, pp 227–275

Fredrickson BL (1998) What good are positive emotions? J Gen Psychol 2:300–319

Friedrich J, Barnes P et al (1999) Psychophysical numbing: when lives are valued less as the lives at risk increase. J Consum Psychol 8:277–299

Frijda NH (1988) The laws of emotion. Am Psychol 43:349–358

Frijda NH, Kuipers P et al (1989) Relations among emotion, appraisal, and emotional action readiness. J Pers Soc Psychol 57(2):212–228

Gleicher F, Petty RE (1992) Expectations of reassurance influence the nature of fear-stimulated attitude change. J Exp Soc Psychol 28:86–100

Hendrickx L, Vlek C et al (1989) Relative importance of scenario information and frequency information in the judgment of risk. Acta Psychol 72:41–63

Higgins ET (1987) Self-discrepancy—a theory relating self and affect. Psychol Rev 94:319–340

Hsee CK (1996) The evaluability hypothesis: an explanation for preference reversals between joint and separate evaluations of alternatives. Organ Behav Hum Decis Process 67(3):247–257

Hsee C, Rottenstreich Y (2004) Music, pandas, and muggers: on the affective psychology of value. J Exp Psychol 133(1):23–30

Isen AM (1997) Positive affect and decision making. In: Goldstein WM, Hogarth RM (eds) Research on judgment and decision making: currents, connections, and controversies. Cambridge University, New York, pp 509–534

Isen AM (2000) Some perspectives on positive affect and self-regulation. Psychol Inq 11(3):184–187

Isen AM, Geva E (1987) The influence of positive affect on acceptable level of risk: the person with a large canoe has a large worry. Organ Behav Hum Decis Process 39:145–154

Jenkins-Smith H (2001) Modeling stigma: an empirical analysis of nuclear images of nevada. In: Flynn J, Slovic P, Kunreuther H (eds) Risk, media and stigma. Earthscan, London, pp 107–131

Jenni KE, Loewenstein G (1997) Explaining the "identifiable victim effect". J Risk Uncertainty 14(3):235–258

Johnson EJ, Tversky A (1983) Affect, generalization, and the perception of risk. J Pers Soc Psychol 45:20–31

Johnson M, Lakoff G (2002) Why cognitive linguistics requires embodied realism. Cogn Linguistics 13(3):245–263

Kahan DM (2008) Two conceptions of emotion in risk regulation. U Penn Law Rev 156:741–766

Kahan DM, Slovic P et al (2007) Affect, values, and nanotechnology risk perceptions: an experimental investigation. George Washington University Legal Studies, Washington

Kahneman D (2000) Experienced utility and objective happiness: a moment-based approach. In: Kahneman D, Tversky A (eds) Choices, values, and frame. Cambridge University Press and the Russell Sage Foundation, New York, pp 673–692

Kahneman D (2003) A perspective on judgment and choice. Am Psychol 58(9):697–720

Kogut T, Ritov I (2005) The "Identifiable Victim" effect: an identified group, or just a single individual? J Behav Decis Mak 18:157–167

Lakoff G, Johnson M (1999) Philosophy in the flesh. Basic Books, New York

Leiserowitz A (2006) Climate change risk perception and policy preferences: the role of affect, imagery, and values. Clim Change 77:45–72

Lerner JS, Keltner D (2000) Beyond valence: toward a model of emotion-specific influences on judgment and choice. Cogn Emotion 14:473–493

Lerner JS, Keltner D (2001) Fear, anger, and risk. J Pers Soc Psychol 81:146–159

Lerner JS, Gonzalez RM et al (2003) Effects of fear and anger on perceived risks of terrorism: a national field experiment. Psychol Sci 14(2):144–150

Loewenstein GF (1987) Anticipation and the valuation of delayed consumption. Econ J 97:666–684

Loewenstein G, Lerner JS (2003) The role of affect in decision making. In: Davidson R, Goldsmith H, Scherer K (eds) Handbook of affective science. Oxford University Press, Oxford, pp 619–642

Loewenstein GF, Weber EU et al (2001) Risk as feelings. Psychol Bull 127(2):267–286

Loomes G, Sugden R (1982) Regret theory: an alternative theory of rational choice under uncertainty. Econ J92:805–824

Mackie DM, Worth LT (1989) Processing deficits and the mediation of positive affect in persuasion. J Pers Soc Psychol 57:27–40

Maia TV, McClelland JL (2004) A reexamination of the evidence for the somatic marker hypothesis: what participants really know in the Iowa gambling ask. Proc Natl Acad Sci 101:16075–16080

Meier BP, Robinson MD (2004) Why the sunny side is up. Psychol Sci 15(4):243–247

Meier BP, Robinson MD et al (2004) Why good guys wear white: automatic inferences about stimulus valence based on color. Psychol Sci 15:82–87

Mellers BA (2000) Choice and the relative pleasure of consequences. Psychol Bull 126(6):910–924

Mowrer OH (1960) Learning theory and behavior. Wiley, New York

Peters E (2006) The functions of affect in the construction of preferences. In: Lichtenstein S, Slovic P (eds) The construction of preference. Cambridge University Press, New York, pp 454–463

Pfister H-R, Bohm G (2008) The multiplicity of emotions: a framework of emotional functions in decision making. Judgem Decis Mak 3(1):5–17

Pham MT (2004) The logic of feeling. J Consum Psychol 14:360–369

Pham MT (2007) Emotion and rationality: a critical review and interpretation of empirical evidence. Rev Gen Psychol 11(2):155–178

Pham MT, Cohen JB et al (2001) Affect monitoring and the primacy of feelings in judgment. J Consum Res 28:167–188

Prior M (1998) Economic valuation and environmental values. Environ Values 7:423–441

Reyna VF (2004) How people make decisions that involve risk. Curr Dir Psychol Sci 13(2):60–66

Reyna VF, Brainerd CJ (1995) Fuzzy-trace theory: an interim synthesis. Learn Individ Differ 7:1–75

Reyna VF, Lloyd FJ et al (2003) Memory, development, and rationality: an integrative theory of judgment and decision-making. In: Schneider S, Shanteau J (eds) Emerging perspectives on decision research. Cambridge University Press, New York, pp 201–245

Roeser S (2006) The role of emotions in judging the moral acceptability of risks. Saf Sci 44:689–700

Roeser S (2010) Emotional reflection about risks. In: Roeser S (ed) Emotions and risky technologies, vol 5. Springer, New York, pp 231–244

Roeser S et al (2009) The relation between cognition and affect in moral judgments about risk. In: Asveld L, Roeser S (eds) The ethics of technological risks. Earthscan, London, pp 181–201

Rottenstreich Y, Hsee C (2001) Money, kisses, and electric shocks: on the affective psychology of risk. Psychol Sci 12(3):185–190

Satterfield T, Slovic S (2004) What's nature worth?. Narrative expressions of environmental values, University of Utah Press, Salt Lake City

Satterfield T, Slovic P et al (2001) Risk lived, stigma experienced. In: Flynn J, Slovic P, Kunreuther H (eds) Risk, media and stigma. Earthscan, London, pp 68–83

Schwarz N (1990) Feelings as information: informational and motivational functions of affective states. In: Sorrentino RM, Higgins ET (eds) Handbook of motivation and cognition: foundations of social behavior. Guilford, New York, pp 527–561

Schwarz N (2004) Metacognitive experiences in consumer judgment and decision making. J Consum Psychol 14:332–348

Schwarz N, Clore GL (1983) Mood, misattribution, and judgments of well-being: information and directive functions of affective states. J Pers Soc Psychol 45:513–523

Schwarz N, Clore GL (1988) How do I feel about it? Informative functions of affective states. In: Fiedler K, Forgas J (eds) Affect, cognition, and social behavior. Hogrefe International, Toronto, pp 44–62

Sloman SA (1996) The empirical case for two systems of reasoning. Psychol Bull 119(1):3–22

Slovic P (ed) (2010) The feeling of risk: new perspectives on risk perception. Earthscan, London

Slovic P, Flynn J et al (1991) Perceived risk, trust, and the politics of nuclear waste. Science 254:1603–1607

Slovic P, Finucane ML et al (2002) The affect heuristic. In: Gilovich T, Griffin D, Kahneman D (eds) Intuitive judgment: heuristics and biases. Cambridge University Press, New York, pp 397–420

Small DA, Loewenstein G (2005) The devil you know: the effects of identifiability on punitiveness. J Behav Decis Mak 18:311–318

Small DA, Loewenstein G et al (2007) Sympathy and callousness: the impact of deliberative thought on donations to identifiable and statistical victims. Organ Behav Hum Decis Process 102:143–153

Stern P, Dietz T (1994) The value basis of environmental concern. J Soc Issues 50(3):65–84

Stevens SS (1975) Psychophysics. Wiley, New York

Sunstein CR (2007) On the divergent American reactions to terrorism and climate change. Colum L Rev 503:503–557

Trepel C, Fox CR et al (2005) Prospect theory on the brain? Toward a cognitive neuroscience of decision under risk. Cogn Brain Res 23:34–50

Vastfjall D, Peters E et al (2008) Affect, risk perception and future optimism after the tsunami disaster. Judgem Decis Mak 3:1

von Neumann J, Morgenstern O (1947) Theory of games and economic behavior. Princeton University Press, Princeton

Weber EU, Hsee CK (1999) Models and mosaics: investigating cross-cultural differences in risk perception and risk preference. Psychon Bull Rev 6(4):611–617

Weber EU, Johnson EJ (2006) Constructing preferences from memory. In: Slovic P, Lichtenstein S (eds) Construction of preferences. Cambridge University Press, New York

Wegener DT, Petty RE (1994) Mood management across affective states: the hedonic contingency hypothesis. J Pers Soc Psychol 66:1034–1048

Wilson RS, Arvai JL (2006) When less is more: how affect influences preferences when comparing low- and high-risk options. J Risk Res 9(2):165–178

Yeung CWM, Wyer RS (2004) Affect, appraisal, and consumer judgment. J Consum Res 31:412–424

Zajonc RB (1980) Feeling and thinking: preferences need no inferences. Am Psychol 35:151–175
Zeelenberg M, van Dijk WW et al (2000) On bad decisions and disconfirmed expectancies: the psychology of regret and disappointment. Cogn Emotion 14:521–541
Zeelenberg M, Nelissen RMA et al (2008) On emotion specificity in decision making: why feeling is for doing. Judgem Decis Mak 3(1):18–27

Chapter 4
Sociology of Risk

Rolf Lidskog and Göran Sundqvist

Abstract Risk is a relatively new object of sociological research, but this research field has grown rapidly over the last three decades. This chapter argues that the task of sociology is to contribute to risk research by emphasizing that risks are always situated in a social context and are necessarily connected to actors' activities. Thus, sociology opposes the reification of risks, where risks are lifted out their social context and are dealt with as something uninfluenced by activities, technologies, and instruments that serve to map them. The chapter comprises four sections, the first being a general introduction. The next section presents a historical perspective on sociology and risk. It starts by briefly describing what sociology is, followed by a discussion of how the concept of risk becomes an object of sociological thought. The section ends by presenting and discussing three different sociological perspectives on risk, all providing different contributions. The third section focuses on current strands of sociological risk research. It starts by giving an overview of three different sociological approaches to risk: Mary Douglas's cultural theory of social order, Ulrich Beck's theory of reflexive modernization and the risk society, and Niklas Luhmann's system theory. All three approaches conceptualize risk differently and make different contributions to the sociological study of risk. The review of these theories is followed by a presentation of five central discussions within the sociology of risk: risk governance, public trust, democracy and risk, the realism–constructivism debate, and governmentality and risk. Finally, the fourth section briefly presents some areas in need of further sociological research.

R. Lidskog (✉)
Örebro University, Örebro, Sweden
e-mail: rolf.lidskog@oru.se

G. Sundqvist
University of Oslo, Oslo, Norway
e-mail: goran.sundqvist@tik.uio.no

S. Roeser et al. (eds.), *Essentials of Risk Theory*, SpringerBriefs in Philosophy,
DOI: 10.1007/978-94-007-5455-3_4, © The Author(s) 2013

Introduction

Over the last few decades, we have witnessed an explosion of risk management practices across a wide range of organizational contexts, such as environment, health, food, crime, media, and traffic (Lupton 1999; Tulloch 1999; Hutter and Power 2005; Hughes et al. 2006; Kemshall 2006; Taylor-Gooby and Zinn 2006; Renn 2008; Reith 2009). All organizations have to relate to risks in their environment. These risks are not only connected with industrial activities, such as harmful substances or technical artifacts. Instead, a growing number of risks concern how actors act upon what they see as risks associated with an organization. Public relations, risk communication, and participatory approaches to risk management have emerged as means to handle diverging interests in society; not least public perceptions could be a source of risk in the sense that these perceptions could pose a threat to the legitimacy and stability of existing ways of managing risk (Power 2007, p. 21). Thus, risk management focuses on how organizations deal not only with the technical calculation of risks, but also with the actors they perceive as possible threats and potential risks to the stability of the organization. Managing such processes is a matter not only of rules for how we should mitigate or accept certain environmental hazards or health risks, but also of rules regarding the process itself, and of activities that target the understanding of risk and deal with public opinion and perceptions concerning it.

This development implies that risk management is no longer limited to a specific sector dealing with certain kinds of risks (such as nuclear power, the chemical industry, and road transport). Risk management has instead become an integral part of managerial language and organizational activities, and all organizations—private companies, governmental agencies, interest organizations, and nongovernmental organizations—have to deal with risks and have made risk management an important rationale for their activities. Not only organizations, but also citizens must include risk thinking when organizing their social world (Tulloch and Lupton 2003; Höijer et al. 2006). Previous certainties—social forms such as nation state, class, ethnicity, traditional family structures, and gender roles—which people use to map out their future are now eroding and citizens have to navigate their lives without them (Beck 2002, p. 22). There is not only public concern about new technologies, but also about how to organize social life and who and what to trust in an uncertain world. This has led researchers to claim that we today face a grand narrative of risk and risk management at the global level (Power 2007, p. viii). Thus, society has no option but to organize itself in the face of risk (Lidskog et al. 2005). Assessing, managing, and communicating risk has become a veritable industry.

This progression from risks associated with certain industrial activities to risks associated with individual and organizational behavior has led to a strong call for sociological analysis. However, scientific development is not only a reflection of changing societal conditions; it is also a driving force of this change. Social theorists have claimed that we today live in a risk society (Beck 1992), a culture of fear (Furedi 2002), and in a social climate that fosters insecurity, fear, and risk

(Giddens 1990; Bauman 2006; Furedi 2008). Citizens and organizations have been provided with a new risk language, causing them to evaluate different phenomena and activities in terms of risk. Thus, there is a dynamic relation between societal development and our understanding of and reflection on this development and society at large.

Within science we have witnessed a development from technical risk analysis—populated by philosophers, statisticians, and economists—to the broader field of risk governance, in which social scientists ponder how actors understand risks as well as how they handle them. Risks are put in specific contexts, which implies a call for social science in general and sociology in particular to develop knowledge on risk. This is the reason why we today can see an explosion of social scientific literature on risk; in particular how to analyze and manage risk.

The contribution of sociology to the field of risk research is mainly that society is differentiated, which means that also cognitions, understandings, and feelings of risk are differentiated. Actors have various cultural belongings and structural positions which make them understand reality differently, and therefore also act differently. Thus, to develop sociological knowledge on risks implies to contextualize risks; they are not the result of a calculation made beyond society, but instead the result of how actors, located in specific social settings, understand and manage certain phenomena. Risk is for sociology always a particular risk situated in a specific context.

Risk is a relative new object of sociological research, and even if this field has grown rapidly over the last three decades, it has not yet fully been institutionalized as a self-evident subfield of sociology (Krimsky and Golding 1992; Zinn 2008, p. 200). In addition, the sociological discipline covers a broad range of perspective and traditions, which is reflected in its research on risks; there are a number of sociological ways to conceptualize, understand, and conduct research on risk. Sociological thought spans from rational choice approaches to cultural theory; it encompasses micro-sociological theories on the construction of self-identities as well as macro-sociological theories on world systems. Thus, it is not an easy task to map out the sociology of risk. We will do so, however, by starting from some central assumptions of sociology, reviewing some of its most well-known approaches to risk, and finally by presenting a few important ongoing discussions and pointing out important areas in need of further research.

The aim of this chapter is to construe a sociology of risk that does not take technical risk analysis as a point of departure, to prevent sociology from being given a too restricted role in researching risk. Instead, we argue that the task of sociology is to contribute to an understanding of the risk field in which risks always are situated in a social context and are necessarily connected to actors' activities. Thus, sociology opposes any kind of reification of risks, in which risks are lifted out of their social context and dealt with as something uninfluenced by the activities, technologies, and instruments that serve to map them.

The essay comprises four sections, this introduction being the first. The next and second section presents a historical perspective on sociology and risk. It starts by briefly describing what sociology is, followed by how the concept of risk gradually becomes an object of sociological thought. The section ends by presenting and discussing three different sociological perspectives on risk, all providing different contributions to the risk research field. The third section focuses on current strands of sociological risk research. It starts by giving an overview of three different sociological approaches to risk: Mary Douglas's cultural theory of social order, Ulrich Beck's theory of reflexive modernization and the risk society, and Niklas Luhmann's system theory. All three approaches conceptualize risk differently and make different contributions to the sociological study of risk. The review of these theories is followed by a presentation of five central, partly overlapping and ongoing discussions within the sociology of risk: risk governance, public trust, democracy and risk, the realism–constructivism debate, and governmentality and risk. Finally, the fourth section, based on these current discussions, briefly presents some areas in need of further sociological research.

History

What is Sociology?

The origin of sociological thought can be traced to the end of the eighteenth century in Western Europe (Eriksson 1993). At this time, questions arose about the social order, the division of labor, social hierarchies, social cohesion, and individualization. A concept of society emerged that did not correspond with the sum of its population, but instead was a social phenomenon *sui generis*, a phenomenon with its own characteristics.

This understanding does not mean that society was external to human beings, but rather was something constitutive of them. The history of human beings and the history of society are two sides of the same coin. This perspective on society emerged with thinkers such as Adam Ferguson (1767), John Millar (1771), and Adam Smith (1776). Whereas Thomas Hobbes (1651) understood man as a "rational wolf", in need of external pressure from outside (in form of norms, laws, and force) to enable social life, these social thinkers understood human beings and society as interdependent. In their view, society was more than an external environment; it was also something inside us. Our words and deeds were not only individual acts; they were also social products. These thinkers were the predecessors of sociology, and 100 years later, the discipline of sociology first saw the light of day.

The growth of sociology is intertwined with the development of empirical data collection and social statistics (Calhoun et al. 2007, pp. 13–18). In a number of European countries, governments had begun to regularly collect information about their populations. Statistical analysis grew strongly, and the national census—originally a way to keep track of adult males' availability for military service—emerged as a regular activity of the state, taking its modern form in the nineteenth century. British Parliamentary investigations of industrial conditions provided the empirical basis for Karl Marx's initial theorizing, empirical data on deaths collected by governments and churches for Émile Durkheim's study of suicide, and publicly gathered data for Max Weber's investigation of German peasants and "junker capitalism".

Sociology was constituted as an empirically based social science, with an emphasis on the importance of context (social, material, economic, cultural) for understanding social life. It also emphasized that society was a social phenomenon in its own right; it was not just an aggregate of individuals. Social practices and collective understandings were not possible to explain by referring to individual human beings.

Phenomena and activities should be understood and explained in relation to their social contexts. And this applies not only to norms and artifacts, but also to actors and knowledge. Hence, sociology is critical of individualistic explanations, though without rejecting the importance of actors. Individual human beings are always situated in social settings, which have to be considered when explaining their cognitions, feelings, and practices.

Risk in Sociology: From Social Problems to Risk

The relationship between individuals and society has been central to sociological thought since its origin (Giddens 1984). There have been—and still are—many ways for sociology to explore and explain this relationship. Even if everyone agreed that there was no such thing as "pure individuals"—human beings unaffected by society whose thoughts, emotions, and wills developed apart from society—they all emphasized that this relationship was not a harmonious one. People find themselves limited by social positions and cultural belongings and struggle to transform structural barriers and cultural restrictions. At the same time, sociologists emphasized that human beings' aspirations and ideals come not only from inside, but also from outside—from social norms and ideals that surround them and which they gradually have internalized. They do not develop their own goals, values, and preferences apart from those that exist in society, but instead develop them in relation to these. Thus, the human being is neither a puppet, nor her own master. Social structures and cultural belongings not only serve as barriers to social action, they also enable it.

In classical sociology, social problems and not risks were the focal point. Its classical thinkers—Karl Marx, Max Weber, Émile Durkheim, and Georg Simmel— were preoccupied with the emergence of modern society, not least the development of industrialization, urbanization, and rationalization and their degrading effects on human beings. Different kind of social problems were put to the fore in the sociological analysis, and different angles were tried in exploring these problems. Risk was not incorporated as a conceptual lens through which these problems were understood and analyzed. Instead, risk research emerged and developed without any relation to sociological thought. The reasons for this were dual: disciplines dealing with risks did not see any relevance of sociological analysis, and sociology did not see risk as a relevant object for sociological research.

Traditional risk concepts have been developed within a framework where risk is technically defined. For *technical risk analysis*, risk means to anticipate potential harm to human beings, cultural artifacts and ecosystems, to average these events over time and space, and to use relative frequencies (observed or modeled) as a means to specify probabilities (Renn 1998, p. 53). Thus, risk concerns a situation or event in which something that human beings value is at stake and where the outcome is uncertain (Jaeger et al. 2001, p. 17).

This kind of analysis implies that one set of experts establishes the probability and magnitude of the hazards and another set of experts evaluates the costs and benefits of various options. Thereafter, political priorities are invoked in order to make decisions on regulating (forbidding, controlling, permitting) certain risks (Amendola 2001). Science is pivotal in measuring and assessing risks, and therefore experts should guide the risk management processes.

Thus, technical risk analysis is an un-sociological understanding of risk; it does not consider the broader social, cultural, and historical context from which risk as a concept derives its meaning (Lupton 1999, p. 1). In response to this kind of analysis, three different sociological perspectives have emerged, all making different sociological contributions to the field of risk research: the social construction of misperception of risk, the social amplification of risk, and the social construction of risk. With this development, important ideas from classical sociology have gradually been taken advantage of and made to influence risk research.

Sociology Explaining Public Misperceptions of Risk

Risk researchers and risk managers gradually recognized that the public's perception of risk was different from the view held by the experts. The nuclear researcher Chauncey Starr's seminal article "Social benefit versus technological risk", published in *Science* in 1969, emphasized the importance of considering public acceptability of risk (Starr 1969). He found that risk tolerance was correlated with a number of social components. For example, the public more easily accepted voluntary and familiar risks than involuntary risks (comparing risks associated with the same level of social benefit).

Social and behavioral scientists have devoted themselves to finding out how different groups and individuals perceive risks (Gutteling and Wiegman 1996; Breakwell 2007). Their point of departure—most explicitly within the *psychometric school of risk analysis*—is that for laypersons risk is a subjective assessment in which contextual factors play an important role. This does not mean that citizens' reasoning is irrational or haphazard. On the contrary, it is possible to find cognitive patterns and trace causal factors that explain citizens' risk perceptions. It is found that factors such as novelty (how new a risk is), dread (how feared the risk is), and if the cause of the risk is seen as tampering with nature, are significant factors shaping risk perception (Slovic 1987; Sjöberg 2000). Perceived influence and power are also important factors, as is cultural belonging (Finucane et al. 2000; Zinn and Taylor-Gooby 2006). Also of importance is the social and spatial context in which people make judgments (Lidskog 1996; Wester-Herber 2004). Much research has also been concerned with how different social groups access, interpret, understand, and respond to different forms of information in diverse contexts (Slovic and Peters 1998; Bickerstaff and Walker 2001; Howel et al. 2002).

Thus, there are a number of contextual and social factors that explain why the public does not assess risk in a similar way as the experts. This perspective considers citizens' perception and understanding of risk, and does not give any attention to the perception of experts and how they understand and measure risks. Instead, the technically defined risk is taken for granted; it is seen as an objective phenomenon in which scientific measurements and statistical calculations give correct, or at least the most valid, knowledge on the character of the risk.

When technically defined risks are not seen as contextually generated, the public's risk understanding is portrayed as biased or incorrect compared to the experts' more accurate assessment. The difference between expert and citizen understanding is interpreted as caused by public ignorance or misunderstanding of science (Irwin and Wynne 1996; Levinson and Thomas 1997). By informing—and sometimes even educating—laypeople about the "real risk", it is believed that the public would correct its judgment and accept risks that experts and regulators have found to be acceptable (Gouldson et al. 2007). According to this view—commonly labeled as *the deficit model of public understanding of science* (Irwin and Wynne 1996)—knowledge is first produced in a closed circle of scientists, after which it should be disseminated to the public (who in many cases are unable to understand science properly). This view is a variant of the "sociology of error", which explains what is seen as error and falsehood in science with reference to contextual factors (such as traditions, ideology, conventions, authority, and interests) and what is seen as true and valid knowledge with reference to observations and reasons. The role of sociology, however, is to explain error, not truth (Bloor 1976).

An implication of this perspective is that risk assessment concerns objective analysis aiming to produce factual knowledge about specific risks, while risk communication is about the distribution/transmission of this factual knowledge to the public. To make risk communication effective, it is important to understand how different segments of the public understand risks, and assess the sources of information, to be able to effectively inform them about risks in different circumstances. The sociological task is to provide knowledge about these factors that result in public misperception of risks.

Sociology Explaining Amplifications of Risk

The deficit model draws a sharp line between risk as defined and assessed by experts and as understood by the public. However, many sociologists seek a more sophisticated way to understand the clash between experts' and the public's understanding of risk. In the late 1980s, the *Social amplification of risk* approach was developed (Kasperson et al. 1988; Pidgeon et al. 2003). This is a communication model according to which an original physical event generates a signal that passes different social stations that amplify or attenuate the signal. The model explains why risks evaluated by technical risk analysis as being similar may receive different levels of attention in society at large.

The basic assumption is that a risk event has certain physical characteristics, such as material damage, injuries, and deaths. These characteristics provide an original signal that is then transformed in the communication process. The risk event itself has no meaning, but the social stations of amplification charge it with meanings and messages. The social amplification of risk explains how risks and risk events interact with psychological, social, institutional, and cultural processes in ways that amplify or attenuate risk perceptions and public concern, and thereby shape risk behavior.

It also explains the development of secondary effects of risk events, that is, risks caused by the amplification processes and not by the signal itself (Kasperson et al. 1988). Social amplification of a risk event associates it with meaning that may result in changed policy regulation, new conditions for insurance, consumer boycotts of a product, decreased institutional confidence, and social stigma. Thus, the amplification may result in social or economic consequences that go far beyond the direct consequences of the risk event.

Thus, technical risk analysis cannot provide information about how risks are amplified in society, because understanding and explaining processes of amplification is solely a task for social and behavioral science. The social amplification approach proposes a division of labor in which technical risk analysis is concerned with investigating the original signal whereas social science in general and sociology in particular analyze how this signal is transformed by society.

The social amplification of risk approach is symmetric in the sense that both the attenuation and intensification of the original risk signal are taken into account. Both technical risk analysis and sociological analysis are needed in order to gain knowledge on risk and its consequences for society. The approach contributes an understanding of why certain hazards and events that experts assess as low risk may receive public attention, whereas other hazards that experts consider more severe receive less attention (Kasperson et al. 2003). By encompassing different factors on different levels, it presents a dynamic and multilayered view on how risk understanding develops. It also aims to link the three leading schools of risk analysis—technical risk analysis, psychometric studies of risk perceptions, and sociocultural studies on risk understandings—into a single framework (Pidgeon et al. 2003).

This bridging effort is ultimately based upon a clear division between a physical world of events and a social world of meanings. The amplification process starts with a physical signal, either in the form of an event (e.g., an earthquake) or the recognition of an adverse effect (such as the discovery of climate change). Thereafter, social factors attribute meaning to it. Risk is thereby conceptualized partly as an objective property of a hazard or risk event, and partly as a social construct (Kasperson 1992, p. 158). According to its proponents, this position avoids the two problems of conceptualizing risk in a totally objectivistic or in a relativistic manner (Renn 2008, p. 39).

The approach links different ways to understand and analyze risk. It is, however, a synthesis based on a linear model in which something external to society is channeled through amplifying stations, resulting in different consequences, and where feedback mechanisms and processes of iteration only take place between the social stations of amplification. The hazard itself and experts' calculation of risk (not least through technical risk analysis) are left outside the approach and are not included in the analysis. In this way, risks—or at least risk events and hazards—are positioned as external to society. Fact finding and sense making are seen as different and discrete spheres of activity, the former populated by technical risk analysts and the latter by various segments of the public. This model does not discuss how the risk (in form of risk events, hazards, or the technical calculation of risk) is constructed, but only how it is amplified. Hence, the bridging ambition also results in a reproduction of the divide between expert and public understandings of risk. Not only risks are amplified in this approach, but also the divide between risk and understandings of risk.

Sociology Explaining Risk

In contrast to sociological studies of the misperception of risk and the amplification of risks, the *social construction of risk* approach includes the role of science and technical risk analysis as topics to investigate. Its starting point is that all risks are socially constructed in the sense that risks always exist in contexts (Wynne 1992a). This means that technical risk analysis and experts' assessments of risks have no privileged position; they are only one of many possible ways to frame, define, and understand risks. Thus, knowledge is intimately related to meaning and actors, which means that no kind of knowledge and no kind of actors should be excluded from sociological analysis. Instead, a symmetrical approach is put forward, where all risks—irrespective of how they are assessed and by whom—are seen as socially constructed. Risks, hazards, and risk events are all sociocultural phenomena in their own right, and should not be seen as unproblematic facts that generate specific signals which laypeople then misunderstand or social stations amplify.

This understanding—that also science's assessment of risk should be seen as construction of risk—has been fuelled by recent developments in society. Science was initially applied to a "given" world of nature, people and society, and scientific skepticism demystified the social and natural worlds. Science's own claim of rationality was itself spared from the application of scientific skepticism. According to Ulrich Beck (1992), a process of "reflexive scientization" is gradually taking place, whereby scientific skepticism is extended to consider the inherent foundations and external consequences of science itself. This demystification opens up new possibilities for questioning science and technical risk analysis. This extension of the scope of rational skepticism means that no scientific statement is "true" in the old sense of there being an unquestionable, eternal truth, where "to know" means to be certain (cf. also Giddens 1990, p. 40, 1994).

The implications of this perspective—risk as a product of social processes—are far-reaching. Not only does it mean that the task of sociology is to analyze how actors—including science and risk experts—frame, define, understand, and manage risks. It also implies that the separation of risk regulation into distinct areas—risk assessment, risk management, risk communication—is incorrect. Values are not solely invoked in the initial process of defining risks that then should be analyzed, evaluated, and regulated by technical risk analysis. Instead, they are an intrinsic part of the risk regulation process, as in the process of developing and validating knowledge. Thus, even if they are presented as separate spheres, they are not discrete activities ordered in a linear process aiming to regulate risk, but instead are dynamically related to each other. The problem is that technical risk analysis' definition of risk is preceded by an implicit framing, which is rarely the subject of discussion, either by citizens or by the risk researchers themselves. This framing provides a very restricted understanding of risks and of actors, behaviors, and processes (Wynne 1992a, 2005). When this framing is naturalized—taken as a pre-given way to understand and conceptualize risk—it restricts the role of sociology to investigate and explain why actors' perceptions and understandings of risk differ from those put forward by technical risk analysts.

The social construction of risk approach has been criticized, not only by technical risk analysts, but also by social scientists. To consider risk as a social construct implies, according to its opponents, a far-reaching relativism where risk bears no relation to a reality beyond human consciousness and cultural values. The social amplification of risk was explicitly developed with the aim of transcending the division between naïve empiricism and far-reaching relativism, as an approach that includes both the need for technical risk analysis and the need for cultural theory (Kasperson et al. 1988). Some researchers, such as Ortwin Renn (2008), argue that it is possible to take advantage of both a more contextual understanding of risk and a more traditional analytical and context-less approach to risk. This is done by defining risk as constituted by both physical/material and social/cultural elements (Kasperson 1992, p. 158; Renn 2008, p. 2). This argumentation rests on the general assumption that it is possible to analytically separate values and evidence, social norms and factual knowledge, deliberation and analysis, but that in practice there is a need for better integration of these analytically distinct entities.

As will be shown below, to understand and analyze risk as a social construct does not necessarily imply a strong relativism, but is based on the assertion that risks are social facts that are irreducible to technical measures. Similarly, its view of knowledge—as contextual, unstable, and sociocultural—does not imply a relativism where all standpoints are given the same cognitive value.

Current Research

A number of sociologists, from somewhat differing standpoints, have emphasised that risk has largely replaced the previous notions of fortune and fate (Beck 1992; Bauman 1993; Luhmann 1993; Giddens 1999). In the past, a lack of certainty was attributed to powers (God, nature, magic) beyond human control, whereas today it is attributed to organizations (such as scientific communities, companies, and nation states). Risk is a factor in human decision making because we cannot gain sufficient knowledge about which possible future will result from our decisions. Furthermore, risk is constituted by the distinction between present reality and future possibilities. Thus, it presupposes that the future is not determined, and that human action shapes the future. As Anthony Giddens (1990, p. 3) puts it:

> Modernity is a risk culture... The concept becomes fundamental to the way both lay actors and technical specialists organise the social world. Under conditions of modernity, the future is continually drawn into the present by means of the reflexive organisation of knowledge environments.

However, to reflect on future consequences of human action is nothing new in history. The decisive difference is that in modern societies almost all aspects of social life are included in these reflections and have become objects of decisions and deliberations; hence thinking in terms of risk assessments is a more or less ubiquitous exercise in everyday life (Callon et al. 2009).

In the following, we will present three important sociological contributions to risk research, all of which treat risk as central for society at large. They take into account the broader historical, social, and cultural contexts from within which risk derives its meaning and resonance. Thereafter, we present five thematic areas that are objects of lively discussion in contemporary risk sociology.

Important Theories

Mary Douglas: Purity and Danger

The British anthropologist Mary Douglas (1921–2007), who has inspired many social scientists in the field of risk research, argues in her book *Purity and Danger* that risks should be understood with reference to the social organization (Douglas 1966). The assessments of risks are responses to problems in the social organization of a specific society, but are also resources for building social order and defending social boundaries. The way risks are viewed reflects the organization of society, including its borders to other societies. What we usually understand as threats coming from outside of society are in fact problems within society.

More specifically, Douglas is interested in how societies assess purity and pollution, which she connects with the overarching concept pair of order and disorder. Purity supports order (both cognitive and social) and pollution is what deviates from and threatens order, and should therefore be condemned. According to Douglas, the separation between purity and pollution, the latter signifying danger, is one of the most fundamental conceptual distinctions in our thinking. This division is, however, relative: "There is no such thing as absolute dirt: it exists in the eye of the beholder" (Douglas 1966, p. 2).

It is important to stress that these definitions are not chosen individually, but in a collective process which is compelling for individuals. Demands for purity are simultaneously requirements for social order, for the survival of society. Douglas argues that we all have norms of order and an associated type of purity to defend, but that these orders and types vary between groups and societies.

In every defense against risks, there is a wish to protect a social order that is considered endangered. Discussions about risks include a desirable norm—a norm of purity—from which the seriousness of the risks can be established. This norm makes it possible to require risk reduction and thereby increases purity. Demands for better risk management imply demands for societal change. A norm of purity contains a vision of a societal order that better corresponds to this norm. Therefore, for sociology, risk should never be seen as something out there, separate from society, but as something produced in and by society.

One implication of this perspective is that our use of the concept of risk reveals who we are. Values and beliefs (including preferences and knowledge)—what Douglas calls *cosmologies*—are viewed as coherent and endogenously derived systems (cultural biases), generated from specific patterns of social relations (Douglas 1978). Such cosmologies support and legitimate social relations. Actions, organizations, and knowledge interact to generate and legitimate social relations. This interplay should not be understood as a unidirectional causal relation but as a reciprocal interaction in which knowledge and society are co-produced (cf. Jasanoff 2004).

One well-known tool for interpreting and explaining such co-production is the *grid–group typology* originally developed by Mary Douglas and her coworkers (Douglas 1982, 1996; cf. Thompson et al. 1990). "Grid" describes the internal structure, how roles and activities are positioned, and "group" the external borders, how the boundary between insiders and outsiders is defined. These two dimensions are fundamental to all cultures, and imply four different cultures: hierarchical, individualistic, egalitarian, and fatalistic.

The *hierarchical culture* is characterized by a stable and regulated internal social order (high grid). Group membership is strong (high group); it is clear to everyone who is a member and who is not. This culture is characterized by formal procedures, rules, routines, timetables, and trust in authorities. *The individualistic culture* is the opposite of the hierarchical. Group membership as well as hierarchy are low (low grid and low group). This is a culture of enterprise characterized by uncertainty and change. Decision making is performed with a minimum of formal procedures, and is based on trust in individual competence. *The egalitarian culture* has a strong boundary to other groups and the outside world (high group). Purity is strived for and outsiders are viewed as threats. The internal differentiation is low (low grid); equality is strived for between positions. *The fatalistic culture* is a residual culture. It is neither individualistic nor collectivistic, but rather cut off (low group). It includes individuals who do what they are told, though without the protection of either social privileges or individual skills (high grid). Those who govern activities and formulate plans for what will happen are always someone else.

Douglas has used this typology to explain the existence and distribution of different risk perceptions (Douglas and Wildavsky 1982; Douglas 1992). The way individuals and groups react to risks reveals their cultural belongings. Is the reaction about *embracing, ignoring, rejecting,* or *adapting* (Douglas 1978)? In an uncertain situation of risk, are the possibilities emphasized or the negative consequences? These questions are relevant for all kinds of issues that people see as risks and dangers; they can concern things like nuclear power and biotechnology, but also EU membership and immigration.

In terms of the grid–group typology, Individualists tend to focus on the possibilities, *embracing* risks as opportunities to exploit for personal profit. Fatalists do not know how to react and tend to ignore the risks. Egalitarians mobilize resistance in order to *reject* and eliminate the risks. Hierarchical people try to assimilate and *adapt* to the risks through regulation and control of risk activities.

Ulrich Beck: The Risk Society and Reflexive Modernization

Ulrich Beck's (1992) *Risk Society: Towards a New Modernity*—originally published in German in 1986—is one of the most influential works of social analysis in recent decades. This book is about the reflexive modernization of industrial society. Beck's underlying thesis is that we are not witnessing the end but the beginning of modernity, a modernity beyond its classical industrial design. This guiding idea is developed from two angles. First Beck focuses on the social transformation from an industrial society with its production of wealth to a risk society with its production of risks and social hazards. The other side comes into view when Beck places the immanent contradictions between modernity and post modernity within the industrial society at the center of discussion. Thus, risk and reflexive modernization are the two—intrinsically interrelated—themes of this book. Beck explicitly states that the aim is to seek "to understand and conceptualize in sociologically inspired and informed thought these insecurities of the contemporary spirit, which it would be both ideologically cynical to deny and dangerous to yield to uncritically" (Beck 1992, p. 10).

Just as modernization in the nineteenth century dissolved the structure of feudal society and produced the industrial society, modernization today is dissolving industrial society and another society is coming into being. This new and coming society is "the risk society", which is a distinct social formation just as the industrial society was. The risk society differs very clearly from the industrial class society in that it focuses on the environmental question and the distribution of risks instead of the social question and the distribution of wealth. In both types of societies, risks are socialized, that is perceived as a product of political decisions and human action; but in contrast to the risk society, the classical industrial society saw risks as manageable side effects of the production of wealth. These risks were legitimated partly with reference to the production of wealth and partly through society's development of precautions and compensation systems.

Risk is defined by Beck (1992, p. 21) as "a systematic way of dealing with hazards and insecurities induced and introduced by modernization itself". The risks and hazards of the risk society are different than in the industrialized society, as they are more widespread and serious. For the first time in history, society involves the political potential for global catastrophes. In the risk society, the relation between wealth production and risk production is reversed. The production of wealth is now overshadowed by the production of risks. The risks produced have lost their delimitations in time and space and consequently can no longer be seen as "latent side effects" afflicting limited localities or groups.

At one level, the distribution of risks adheres to the class pattern, but it does so inversely: wealth accumulates at the top, while risks accumulate at the bottom. Therefore, the risk society could be seen as simply strengthening the class society. However, on another level, this is not true. Today's diffusion and globalization of risks entails "an end of the other"; that is private escape routes shrink (it is impossible to buy yourself free from risks) as do the possibilities for compensation. Thus, risk positions are no longer pure reflection of class positions, but instead they transform and replace class positions. One example of this is that property (such as forests) today is being devaluated; it is undergoing a creeping "ecological expropriation" which implies the emergence of new conflicts between the different interests of profit and property. This means that the central conflicts in the future will be not between East and West, between communism and capitalism, but between countries, regions, and groups involved in primary and in reflexive modernization, the latter being those that are striving to relativize and reform the project of modernity.

Global risks—mega-hazards, to use Beck's term—overlap with social, biographical and cultural risks, as well as insecurities. Today, these latter forms of risk have reshaped the inner social structure of industrial society and its fundamental certainties of life: social classes, familial forms, gender status, marriage, parenthood, and occupations. This comprises the other part of Beck's discussion of reflexive modernization.

The theory of modernization is formulated by Beck as the unleashed process of modernization overrunning and overcoming its own "coordinate system". This coordinate system has fixed the understanding of the separation of nature and society, the understanding of science and technology, and the cultural reality of social class. It features a stable mapping of the axes between which the life of its people is suspended—family and occupation. It assumes a certain distribution and separation of democratically legitimated politics on the one hand, and the "sub politics" of business, science, and technology on the other.

Today, a social transformation is underway within modernity, in the course of which people will be set free from the social forms of industrial society. This reflexive modernization dissolves the traditional parameters of the industrial society (such as class and gender). This "detraditionalization" occurs in a social surge of individualization, through which a capitalism with individualized social inequality is developing. Here the family is replaced by the individual as the reproductive unit of the social in the life world.

Having discussed this theory of individualization, Beck turns to the role of science and politics in the era of reflexive modernization. He argues that when encountering the conditions of a highly developed democracy and well-established scientization, reflexive modernization leads to an unbinding of science and politics. Earlier monopolies of knowledge and political action are then differentiated.

In discussing science, Beck makes a distinction between primary and reflexive scientization, with the former meaning that science is applied to a "given" world of nature, people and society, and the latter that the scope of scientific skepticism is extended to encompass the inherent foundations and external consequences of science itself. Reflexive scientization thus entails a demystification and demonopolization of scientific knowledge claims. At the same time, the role of science in the risk society is growing. Today's threats are beyond human perception and experience and it is through science that risks become known. Science thus comprises the "sensory organs" for the perception of today's risks. Taken together, these two parallel and different developments do not mean that science has come to an end; on the contrary, it pervades all areas of modern life. Today's science is undergoing a situation of being dethroned similar to that which happened to (institutionalized) religion. In Beck's secularization model of modern science, the future will bring about a pluralization and a marketization of science.

Beck has been of pivotal importance for sociology, not least by paving the way to make risk a central concern for general sociology. In the wake of *Risk Society*, he has published a number of books in which he further develops his perspective: *Ecological Politics in an Age of Risk* (1995), *Ecological Enlightenment* (1995), and *World Risk Society* (1999). In his later writings—such as *What Is Globalization?* (2000), *Individualization* (2002, with Elisabeth Beck- Gernsheim), *Cosmopolitan Vision* (2006), *Power in the Global Age* (2006), *The Brave New World of Work* (2010), and *A God of One's Own* (2010)—Beck puts more emphasis on reflexive modernization and its importance for all aspects of society such as family, work, religion, and global politics. However, in all his books, irrespective of their subject, the main theme is reflexive modernization and the future development of society.

Niklas Luhmann: System Theory and Risk

Drawing on Talcott Parsons's social theory, the German sociologist Niklas Luhmann (1927–1998) developed a general theory of modern society. The starting point is that there is a fundamental distinction between system and environment and communication is the basic social operation (Luhmann 1984, p. 47). A higher complexity in the environment entails a greater importance for the system to reduce this complexity, otherwise the system will not be operational.

This reduction of complexity is accomplished through functional differentiation, which means that different subsystems develop, each with distinct forms of communication (programs and binary codes). These subsystems are self-referential and auto poetic, which means that their internal orders guide their observations and interpretations and that they are not formed and structured by any external factors.

A subsystem is cognitively open; it is receptive to signals from its environment. At the same time, it is operationally closed. Signals are always transformed into communication through a particular binary code of the subsystem. These codes are abstract and universally applicable distinctions. Science codes a signal in terms of truth/untruth; economy in terms of property/no property; law in terms of legal/illegal; religion in terms of transcendent/immanent; and politics in terms of political power or lack of power and so forth. Luhmann (1989, p. 18) states that:

> The system introduces its own *distinctions* and, with their help, grasps the states and events that appear to it as *information*. Information is thus a purely system-internal quality. There is no transference of information from the environment into the system. The environment remains what it is.

This does not mean that nothing else exists than social systems and their communicative processes. What it says is that external facts can only be taken into account as part of the system's environment and only be understood through communication. There is no position available outside the system; these facts can only be understood from within (Luhmann 1993, p. 5). It is, however, possible to observe the border of a system, and this is done through "second-order observation" (Luhmann 1993, p. 223). First-order observations identify facts and objects as givens, and do not reflect on the distinction used in the observation. Second-order observation is an observation of the distinction implicitly used in the first-order observation; it recognizes which distinction is applied in observing a fact or object. Luhmann (1993, p. 227) stresses that second-order observation is not more true or objective than first-order observation. What second-order observation reveals is that there are no objective facts outside the operation of each subsystem. What may seem like an objective fact in the first-order observation (which takes for granted its own distinction), is a product of a particular distinction made in the observation process. Thus, the same event is coded differently by different subsystems.

For instance, the signal from the Tohoku earthquake in Japan, March 11, 2011, that resulted in more then 15,000 deaths and a nuclear disaster at Fukushima nuclear power plant is coded radically differently by different subsystems. The economic subsystem focuses on price mechanisms, economic compensation, and falling prices of shares; the political subsystem on political legitimacy of decisions concerning the location of the nuclear power plants and how the disasters were handled by authorities, but also on the legitimacy of the political representatives that had permitted this activity; the legal subsystems on violations against the given permissions for the plant and the liability of the company as well as political institutions; science on health consequences of radiation exposure for workers and the local population. It is what is communicated that counts. A phenomenon, an event, or an activity can never in itself create a response; it needs to be subject of communication.

> But as physical, chemical, or biological facts, they create no social resonance as long as they are not the subject of communication. Fish or humans may die because swimming in the seas and rivers has become unhealthy. The oil pumps may run dry and the average climatic temperature may rise or fall. As long as this is not the subject of communication, it has no social effect. Society is an environmentally sensitive (open) but operatively closed system. Its sole mode of observation is communication. It is limited to communicating meaningfully and regulating this communication through communication (Luhmann 1989, pp. 28–29).

Risk is inherently linked to a functionally differentiated society. In contrast to many other theories, Luhmann does not see risk as a result of detrimental activities or as caused by industrial society. Instead, risk is attributed to decision making that may result in negative consequences. Contingency is a central concept for Luhmann, which means that a situation includes a large number of possibilities. To be able to act, it is necessary to choose among these possibilities, but there is no fundamental point—external authority—that tells what to select among the various alternatives. This has to do with the development of the functionally differentiated society.

In contrast to earlier societies, there is no privileged function system in society, which means that a functionally differentiated society has no center. Each subsystem can only refer to its own communication, and only internally can it refer to its environment. Through internal differentiation, the system develops a richer way to manage the complexity of the environment. At the same time, this differentiation results in a higher internal complexity, with different functional subsystems existing side by side and communicating with their own specific codes and with no external authority. Earlier societies' external references (such as religion) are replaced by the social system's self-references in the form of subsystems.

Luhmann defines *risk* as an attribution of an undesired event or possible future loss. Thus, risk is an intrinsic part of a functionally differentiated society. Decisions have to be made without any certainty about what consequences they will lead to. The cause of the damage could either be attributed to the system itself (risk) or something external to the system (danger) (Luhmann 1993, pp. 101–102). This means that risks concern attribution, which becomes even more clear when Luhmann discusses another distinction, namely, between those who take the decision and those who are exposed to its consequences (Luhmann 1993, pp. 105). Those who take the decision face a risk; whereas those who are victims face a danger, that is, those who perceive themselves as exposed to something that they cannot control. Uncertainty is intrinsic to both risk and danger; the difference lies in who is seen to be a decision maker.

Luhmann (1993, p. 109) stresses that "one man's risk is another man's danger" and claims that there is a growing gap between those who participate in decision making and those who are excluded from decision-making processes but have to bear the consequences of these decisions.

Luhmann's system theory provides an alternative understanding, not only of risk but also of society at large. It sees risk as a matter of attribution and communication, associates it with decision making, sees it as inherently linked with a functionally differentiated society, and strongly emphasizes the distinction between risk and dangers and between decision makers and those affected; and in doing so the theory has both received support and met with criticism (Japp and Kusche 2008, pp. 101–103).

Thematic Areas

There is today an ongoing and lively discussion within the sociology of risk. In what follows, we present five partly overlapping areas of central importance in contemporary risk sociology: risk governance, public trust, democracy and risk, the realism–constructivism debate, and governmentality and risk.

Organizational Risk: From Risk Analysis to Risk Governance

To conceptualize an object as a risk entails seeing it as manageable and governable (Baldwin and Cave 1999; Hood et al. 2001; Hutter 2001; Lidskog et al. 2009). Risk creates space for action as it opens the future for calculation, deliberation, and decision making. In this sense, regulation "enrolls" futures and shapes policy formulations (Wynne 1996).

Risk regulation is not only about how to govern an existing reality, it also concerns the transformation of this reality, for instance, by dealing with novel forms of knowledge that have not yet been put into industrial practice (cf. Stehr 2005). Regulation does not only concern how to regulate an existing activity, but also how novel knowledge should be deployed and employed.

As shown above, there has been far-reaching criticism of technical definitions of risk, not least concerning the difficulty of upholding a sharp separation between an objective measure of risk and a sociocultural understanding of risk (Hilgartner 1992; Rosa 1998; Amendola 2001; Todt 2003). The critique was initially directed at problems *within risk* analysis, and consequently public perception of risk was seen as a challenge to how risk was defined and approached by technical and calculative means. In the 1990s, however, there was a shift from this internal focus to the broader question of the legitimacy of government. Organizations must ponder not only how to deal with technically defined risks, but also how to deal with actors who may question both the legitimacy of current methods for regulating risk and the trustworthiness of organizations responsible for this regulation.

Prompted by several regulatory failures, authorities and companies have started to account for and deal with public opinion and public perceptions of risks, not only to handle criticism but also to forestall it (Löfstedt 2005). Programs for risk communication, public relations, stakeholder dialogue, and public involvement are today integral to both public and corporative governance (Gouldson et al. 2007). Calls for more inclusive and transparent processes, public dialogue, and democratic engagement are widespread in society (Irwin 2006). A heightened concern for stakeholder involvement and public inclusion can be seen as a strategy to influence perception, shape understandings, and produce legitimacy.

Michael Power describes this shift as a move from risk analysis to risk governance. The "governing gaze" has shifted from how risk is defined, analyzed, and calculated to the governance of the organizations that analyze risk (Power 2007, p. 19). Even though the call for a more inclusive risk analysis and risk management may have evoked some response from regulatory agencies, it has not directly led to more inclusive and deliberative risk regulation processes. Rather, the shift from risk analysis to risk governance has increased the awareness of how organizations deal with public opinion and public perceptions as a source of risk in the sense that such perceptions could pose a threat to the legitimacy and stability of existing ways of governing risk (Power 2007, p. 21). This justifies research on how organizations deal not only with technically defined risks but also with the actors they perceive as possible threats and potential risks to the stability of the organization.

Risk regulation does not only concern what is acceptable in terms of how we should mitigate or accept certain risks and hazards, but also rules regarding the process itself and activities that target the understanding of risk and deal with public opinion and perceptions concerning it. Thus, risk governance is not limited to technical calculation of risk, but also includes the evaluation of organizational aspects in regulation of risk.

With a focus on risk governance, that is how uncertainties are organized in order to transform them into governable risk, the questions of who should be involved or excluded in risk regulation processes, on what grounds, and what aspects should be made open and transparent to others, gain greater relevance due to the legitimacy gains and losses such decisions may generate. A heightened concern for public involvement in regulation can thus be seen as "a strategy to govern unruly perceptions and to maintain the production of legitimacy in the face of these perceptions" (Power 2007, p. 21).

Public Trust: The Relation Between Experts and Laypeople

Technical risk analysis builds on a sharp boundary between experts and laypeople. Laypeople do not have access to all the knowledge possessed by experts and therefore draw different conclusions about risks, their ordinariness, magnitudes, and impact. In technical risk analysis, scientific knowledge is the norm and this is what experts have but laypeople lack. This difference is what motivates the concept of lay knowledge, of not being an expert. Focusing on what laypeople lack constitutes the basis for *the deficit model* mentioned earlier in this essay (Irwin and Wynne 1996). According to this model, the solution is to inform and educate laypeople in order to give them the capacity to gain correct knowledge and thereby arrive at the same conclusions as experts. Risk psychology and risk communication originally developed as academic fields with the aim to understand how laypeople reason about and assess risks in order to learn how to effectively communicate correct knowledge to this group.

The deficit model has been heavily criticized, not least by researchers in the field of science and technology studies (STS) (Wynne 1995; Irwin and Wynne 1996). These scholars argue that the most important problem is not that the public is unaware of research results and scientific facts, but that scientific experts are unaware of and disinterested in lay knowledge and how laypeople assess the situation when decisions are to be taken on complicated risk issues. Consequently, the problem is not that laypeople lack knowledge or lack trust in expertise, but that experts in technical risk analysis do not trust laypeople. Laypeople have the competence to contribute to discussions and decisions on risks, since these concern much more than scientific facts. If grasping scientific details becomes the most important requirement for participation in risk discussions, the relevance of scientific knowledge becomes heavily exaggerated (Irwin and Michael 2003, pp. 22–28). Despite their lack of scientific knowledge, laypeople are competent actors with developed abilities to reflect on what types and sources of knowledge are of relevance to both risk analysis and risk assessment and why some experts should be more trusted than others.

Today, public involvement is often devoted much attention, but there is a tendency to frame this involvement from a technocratic understanding based on the deficit model (Irwin 2006; Lidskog 2008). In this way, broadened participation gives experts further possibilities to inform the public with the aim of winning acceptance for already proposed decisions. This instrumental ambition can be found in every participatory project, because there are always groups who strive for a specific outcome of the process. Studies have found that when laypeople are not considered competent to influence the decisions—when they are taught instead of listened to—the result is often one of alienation rather than engagement amongst the public (Wynne 2001).

Problems arise when such an overconfident and self-sufficient expert culture tries to communicate the benefits of a risk project. This culture is not interested in reflecting on its own shortcomings, and criticisms from laypeople are understood as based on their not understanding what is best for them (Wynne 2001, p. 447). If expert cultures wish to increase their legitimacy and appear as trustworthy to the public, they should instead be less confident about their own results and open to acknowledging their own limitations.

It is, however, not only uncertainties to which attention should be given attention, but also ignorance. Scientific knowledge is strongly specialized with a narrow focus, which implies that complexities are reduced and alternatives are actively deleted. In order to increase the trustworthiness of scientific knowledge, it is important to make visible the conditions of scientific knowledge production and risk assessment. The implication here is that scientific knowledge alone is not enough when deciding about complicated risk issues. It must therefore be enriched by other types of knowledge, as well as by other perspectives in order to give a more complete and nuanced view of the risks at stake.

The alternative to the deficit model and a technocratic framing of risks is to include a broader understanding of participatory processes, one which acknowledges that other actors than scientific ones can contribute knowledge on risk and therefore should be given possibilities to influence the regulation of risks (Funtowicz and Ravetz 1990; Lidskog and Sundqvist 2011). Hence, questions concerning who formulates the issue, sets the agenda, and exercises power become important. However, this does not mean to replace blind trust in experts with blind trust in laypeople. To draw clear boundaries between experts and lay people and grant one of these priority over the other is not the right way to proceed. Instead it means that the public *always* has important contributions to make in technical discussions on risk issues. These contributions should never be evaluated from the deficit model for the reason that they are not about scientific facts, but about how to assess the relevance, trustworthiness, and ignorance of scientific facts. This kind of public competence, which emphasizes the contextual dimension of science, can always enrich scientific knowledge (Wynne 1993, p. 328).

The public can also contribute knowledge and insights about what they are worried about (Marres 2007; Sundqvist and Elam 2010; Lidskog 2011). This knowledge as well as the assessments made by members of the public are anchored in their livelihoods. Experts and authorities are often completely unaware of the reasons for ordinary people's worries, and it is therefore of great importance to involve concerned groups in decision processes in order to include relevant experiences. Experts and decision makers need to improve their awareness of how worries are the driving force of public engagement and that scientific knowledge rarely is an adequate response to these worries.

Risk and Democracy: The Importance of Framing

How risks are defined is a central topic for sociology to study. The reason for this is that it determines what groups and what competences are considered relevant for taking a stand and making decisions (Lidskog et al. 2011). Sociology contributes to risk analysis by showing how definitions of risks shape social relations and distribute powers to groups, at the same time as other groups are excluded from decision making. In a risk context, a scientific definition of the issue at stake is often assumed. However, such a restriction may lead to reductionism, giving experts too much power, while rendering other important factors invisible. Laypeople are reduced to passive receivers of information who only can contribute their trust and consent regarding expert proposals (Wynne 1992b, 2001).

Sociological studies of expert work do not conclude that risk issues should be handled without experts. What is claimed is that experts alone should not define, investigate, and give answers to risk issues. Risks are too complicated to be delegated to experts. Sociological studies of the social dynamics in definition of risks are valuable for making risk management more relevant and robust. Not least to show how definitions and processes of framing are made and what consequences these processes have for different groups' possibilities to participate in and influence risk management.

Regulating a risk is not only about setting limits, but also about framing the risk as such, deciding what actors are of relevance and should be included in the decision process, what roles, mandates and responsibilities are given to them, and finally and most importantly what to make decisions about (Lidskog et al. 2009). Sociologists and other social scientists have critically scrutinized framing processes in public decision making, and a key finding is that frames concerning technical issues are usually dominated and influenced by experts in a technocratic way (Wynne 2001, 2005). By narrowing the issue, scientific experts exaggerate the scope, power, and importance of scientific knowledge in the public domain, neglecting cultural factors and ignoring citizen competence (Wynne 1992b, 2001; Jasanoff 2010). Paradoxically, science is often accorded the most prominent role even when public dialogue is striven for. A reason for this is that many issues are technically framed, with questions of risk, safety, and effectiveness placed at the center (Wynne 2005, 2010). Scientific expertise is needed to answer such questions, since experts have the resources and competence to know the "true" nature of the issue at stake.

The result is that what is presented as a democratic risk decision process is often a technocratically framed process based on a scientific definition of the risk. The public are invited to participate in a process that is often presented as being about dialogue but the possibility to influence the process according to their own perspectives is often unclear, and in practice very restricted. These processes do not open up decision making to wider evaluation and influence, but instead function to gain legitimacy and acceptance for already defined—and many times in practice already decided—expert-based proposals.

A technocratic framing reduces the role of citizens to one of trusting or distrusting experts; to saying yes or no to already decided proposals and to being restricted to only discussing the local and concrete aspects of a project. Instead, they should be provided with opportunities to define and frame the project in their own way, putting forward what risks they see as relevant and worthy of attention. Sociologists have argued that the discussion should not only include the meaning of the project and its risk from the perspective of the experts and regulators, but also from the perspective of the public and other stakeholders (Gieryn 1999; Hilgartner 2000; Irwin and Michael 2003; Jasanoff 2005; Wynne 2005).

Since there is no correct framing of risk issues, risk management will always be surrounded by conflicts. Different groups frame problems differently and give them different priorities. Sometimes it is the case that what one group considers to be a solution to a problem, another group considers to be part of the problem. Some suggest that nuclear power is a sustainable energy solution, because it does not lead to carbon emissions, while others argue that the radioactive waste makes it anything but a sustainable and secure long-term source of energy. In this situation, the option of denying the existence of different frames is a dead end. Instead, the first step toward a robust solution is to acknowledge the existing frames and welcome different groups to contribute their own perspectives, using their own frames, knowledge, and values. Frequently, this opening up of the framing is met with criticism and opposition from those groups that are already handling the issue from a particular frame. They have invested time, money, and prestige in the project and are therefore reluctant to change the established framing of the issue and its particular way of handling it. But taking public involvement seriously entails a more democratic framing of issues, in the sense that issues have to be connected to public concerns (Marres 2007). If we are to find a way beyond pendulum swings between technocracy and populism—that either scientists or laypeople should decide—various groups of frames must meet on a more equal footing.

Produced Risk: Beyond Realism and Constructivism

An ongoing controversy in risk research is between realism and social constructivism. Do risks possess physical characteristics that exist independently of cultural and social contexts, including actors' perceptions, or are they socially and culturally constructed attributes, produced and shaped by these contexts? Technical risk analysis is based on realism, and sees risks as independent of their context. As described earlier in this essay, many social scientists and sociologists accept technical risk analysis and its realism, but add studies on why the public accepts some risk analyses and rejects others. The public's risk assessments are understood as social constructs, whereas experts' risk assessments are seen as realistic descriptions. But among social scientists, we also find those who question the realistic approach and consider it wrong. Instead, they want to go beyond this dichotomy in order to find a middle ground between realism and social constructivism. Many scholars claim that there is a third way between "naïve realism", and "idealism" (Renn 2008), positivistic and constructivist paradigms (Rosa 1998), and "pure realism" and "radical constructivism" (Zinn 2008). The first implies the existence of an empirical and objective reality outside human perception, and the latter a subjective and cultural understanding shaped by humans and with no necessary connection to an objective reality.

However, the quest to find a middle way, or third way, between subjective and objective reality—between something internal and something external to human beings and society—reproduces what it tries to transcend. The first is based on causal laws of material reality, and the second, on a social world of opinions and norms. This point of departure is questioned by certain sociologists, claiming that the focus should instead be on the dynamic interplay between different factors that make up reality (Irwin and Michael 2003; Latour 1993, 2004, 2005). Reality is neither reducible to something out there, beyond human action, nor reducible to something in there, to human thoughts and actions. Instead it is co-produced by many factors.

However, social constructivism has been and still is an important tradition within the sociology of risk. Its historical roots go back to classical sociology, not least the work of Émile Durkheim (1858–1917). Durkheim elaborated a unique domain for sociology by demarcating a social reality totally different from that of biology and psychology. "The social"—or social facts, as he called it—is a reality in its own right, irreducible to other levels of reality (Durkheim 1982). The task of sociology was to explain social facts, and these explanations should not include any findings or factors from the psychological (individual) level or the biological level. The result was a specialization and division of labor among academic disciplines, where every discipline has its own domain and unique explanations. This understanding of sociology entailed that everything that exists outside the social domain was disregarded. The social domain was considered autonomous with regard to other domains.

The implication was that when analyzing risk, sociology should study people's interpretations and experiences of these risks, and how these are bound to social structures that steer perceptions and actions. Material objects and technical artifacts were left outside the analysis, and seen as not having any power to influence what is taking place in the social domain. Experiences of risks should not be explained with reference to nature or artifacts, but only social factors. For example, when sociology explains people's worries about nuclear waste, the focus should not be on the strength of the canisters as a technical barrier to protect the biosphere from radioactivity, but on people's opinions about these barriers and how these influence their assessment of radioactive risks. Thus, the objects of sociological analysis are perceptions, interpretations, and socializations to social patterns of risk attitudes toward radioactivity and disposal of nuclear waste.

This sociological purification of a social dimension has been successful in so far as it has created a distinct niche for sociological thought and provided important knowledge concerning how people perceive, understand, and act upon risks. Nevertheless, its strong separation between nature and society, with sociology only investigating the latter, is problematic. This strong focus on the social dimension has led to a paradoxical understanding of nature and artifacts, which Bruno Latour (2004, p. 33) has aptly described as follows:

> Those who are proud of being social scientists because they are not naive enough to believe in the existence of an "immediate access" to nature always recognize that there is the human history of nature on the one hand, and on the other, the natural nonhistory of nature, made up of electrons, particles, raw, causal, objective things, completely indifferent to the first list.

The consequence of social constructivism is that we, on the one hand, find a society with a history and on the other a nature without history. Latour is critical of this kind of approach, which leaves important aspects of reality outside sociological analysis. According to him, the task of sociology is to transcend both realism and social constructivism, a task that necessarily entails that dichotomies—such as those between nature and culture, social and technical, actor and structure, science and society—be critically studied and not taken for granted. How and why these dichotomies are produced and reproduced should also be explained.

Latour's proposal for transcending the dichotomy between realism and social constructivism is to focus on the *production of risk*. Risks are produced by practices, by actors using instruments and technologies. It is therefore misleading as a sociologist to focus on perceptions, opinions, and experience. Instead, the focal point for sociology should be to explore how risks are produced, by what means, and with what effects. The focus on practices means that there is no "real risk" behind our perceptions and actions. There are no risks separate from actors and society, possible to observe by actors. Instead, there are a number of actors and activities where nature, technology, and culture interact, resulting in the production of risks. There are no risks beyond socially produced risks, that is, beyond the measuring and monitoring of risk. Through these practices not only knowledge about risks is produced but also the risks as such. Thus, practices are performative; they not only describe reality but also shape it. By studying these practices, sociology can transcend the dichotomy between realism and social constructivism.

Governmentality: Toward an Individualized Risk Management

Ulrich Beck emphasizes that the current society is increasingly individualized, in the sense that individuals are seen as being responsible creators of their own lives and are therefore constantly required to make their own decisions. "The choosing, deciding, shaping human being who aspires to be the author of his or her own life, the creator of an individual identity, is the central character of our time", as Beck (2002, p. 23) puts it.

This individualization, however, does not necessarily mean the achievement of greater personal freedom. Beck grasps this development with the term "institutionalized individualism" (Beck and Beck-Gernsheim 2002). At the same time, as nation-states have outsourced many of their functions and operations, there is an insourcing of functions to the individual level. What the nation-state, the employer, the union, or the family once provided is now presented as being the responsibility of the individual. Thus, individualization in this sense does not mean freedom of choice, but instead the compulsion to choose in a situation where no certainties exist. It is a "precarious freedom" centered on imperatives such as think, calculate, plan, adjust, negotiate, define, and revoke (Beck 2002). But even though we often lack knowledge of what choices are best, it is demanded of us to make individual decisions and be responsible for the consequences.

There is a tension between institutionalized individualism and the risk society thesis about mega-hazards beyond human control. According to Beck (2008), the government of incalculable risks and mega-hazards leads to the irony of putting an end to the free liberal society in the ambition of protecting citizens from risks. At the same time, individuals are continuously ascribed responsibility for risks that are impossible for them to manage.

However, a certain strand of sociological thought, following Michel Foucault's (1991) work on governmentality, cultivates a perspective that takes neither individualism nor the character of risks and the risk society for granted. Instead they argue that risks should be conceptualized and understood as a way of steering practice. The task of sociology is to study how, through technical apparatus and administrative institutions, incalculable dangers are made into knowable and governable risks. Risks become a way of ordering reality and making it calculable, and expert knowledge is decisive in this (Rose 1993; Dean 1999).

Instead of making use of coercive power, the government can steer through norms, knowledge, and individual self-discipline. The reason for this is that we have today an "advanced liberal society",' based on a clear division between the state and the civil society (Rose 1996, 1999). Civil society has emerged as an autonomous sphere in which individuals can express themselves as free citizens. In protecting this autonomy, coercive means of governmental control are precluded, which means that more sophisticated instruments and mechanisms need to be developed, technologies for governing at a distance (Rose and Miller 1992).

This way of exercise power, in which those who are controlled feel autonomous, is based on tools for the self-development of those who are governed. The responsibility is placed on citizens to govern themselves, to act upon themselves, and be responsible "for the security of their property and their persons, and that of their families" (Rose 1999, p. 247). An almost paradoxical relationship is created between the state and the civil society, in which the exercise of power is conducted with the goal of not being visible. It is characterized more by bringing citizens to perform a regulated freedom than by imposing on them coercive measures (Rose and Miller 1992, p. 174).

The strong emphasis on individuals as being responsible governors of their own lives creates dilemmas. Increasingly, individuals have to face and make decisions on a range of issues characterized by uncertainty. An example of this is how genetic risks are governed with the aim to improve the quality of the population. During the development of the welfare state, it became an important task for the government and the public administration to guide, control, and intervene in the reproduction of the population, but today these decisions are delegated to individual citizens. The problem is no longer framed as improving the quality of the population but as a question of individual lifestyles. Today, reproduction is about promoting the self-governance of the client (Novas and Rose 2000). The responsibility to govern genetic risks—to decide about having children and informing others about one's own genetic risks—has been made into a lifestyle choice. However, plenty of experts are willing to give advice on how to make your own lifestyle possible, and guide your choice in certain directions.

Risks thereby constitute a strategy for disciplinary power to monitor and govern individuals and thereby whole populations (O'Malley 2008). Those individuals that deviate from what is presented as normal behavior are seen as "at risk", and need to be controlled with the aim of achieving behavioral modification. This control is primarily that of self-management, with individuals being urged to protect themselves from certain risks (Giddens 1991). Risks are thereby de-socialized, privatized, and individualized; they become a responsibility of the individual, and a way for government to govern the conduct of individuals. The sociology of risk should therefore be devoted to studying questions about how problems are defined, by whom and in relation to what goals, and through which practices, technologies and rationalities this governing is accomplished and authority exercised.

Further Research

As emphasized in the introduction, the specific contribution of sociology of risk is to place risk in its social context. There are no risks "out there" in the sense of being independent of the society in which they emerge, are measured and monitored. Society is differentiated, which means that cognitions, understandings, and feelings of risks are differentiated. Actors—including scientific ones—have various structural positions and cultural belongings and therefore understand risks differently. To develop sociological knowledge on risks implies to contextualize risks; to associate them with specific actors, institutions, and settings. This means that no conceptualization, regulation, or research on risks is beyond sociological exploration; and furthermore, scientific definitions of risks and technical risk analysis should be proper study objects for sociological investigation.

This does not imply a reductionism and relativism, seeing different actors' understandings of risks as all that exists. On the contrary, risks should be understood as produced through social activities where nature, technology, and culture interact. Sociology of risk should not be restricted to investigating risk perceptions, but should also study definitions and usage of risks, including how different actors deal with risky nature and unruly technologies; how these are framed and regulated, and as a consequence of these activities, produced.

The five thematic areas described above have by no means been given a final answer, but are in need of further research. As already emphasized, these areas are interrelated; organizational aspects of governing risks, public inclusion in risk regulation, framing and production of risks, and the monitoring of individuals' risk behavior are interconnected. As with many other disciplines, sociology consists of different theoretical traditions, methodological assumptions, and analytical approaches. Therefore, it will never be able to give simple, single, and final answers to complex issues, and the sociology of risk is no exception of this. It is, however, able to gain knowledge on important topics, and while not producing final knowledge at least it may produce more and better knowledge—theoretically informed and empirically sensitive—on the function and place of risks in different social settings.

Studying processes of risk assessment and risk management, the sociology of risk could make important contributions to preparing and realizing a political and democratic discussion on risk issues, controversial as well as uncontroversial. By identifying and clarifying the political aspects of these objects, making frames and framing processes visible, and showing how technologies and political devices are embedded in social processes, it opens up risk regulatory processes for public scrutiny and evaluation. What may originally be framed as technical issues, only relevant for a specialized group of experts, will thereby become relevant for citizens. Risk regulation is about more than just choosing the best regulatory instruments and finding the best technical solutions to predefined risks. It concerns building society and choosing a future.

References

Amendola A (2001) Recent paradigms for risk informed decision making. Safety Sci 40(1):17–30

Baldwin R, Cave M (1999) Understanding regulation. Oxford University Press, Oxford

Bauman Z (1993) Postmodern ethics. Blackwell, Oxford

Bauman Z (2006) Liquid fear. Polity, Cambridge

Beck U (1992) Risk society. Towards a new modernity. Sage, London

Beck U (2002) A life of one's own in a runaway world: individualization, globalization and politics. In: Beck U, Beck-Gernsheim E (eds) Individualization, institutionalized individualism and its social and political consequences. Sage, London, pp 22–29

Beck U (2008) Living in the world risk society. Econ Soc 35:329–345

Beck U, Beck-Gernsheim E (2002) Losing the traditional. Individualization and "precarious freedom". In: Beck U, Beck-Gernsheim E (eds) Individualization. Institutionalized individualism and its social and political consequences. Sage, London, pp 1–21

Bickerstaff K, Walker G (2001) Public understandings of air pollution: the "localization" of environmental risk. Glob Environ Chang 11:133–145

Bloor D (1976) Knowledge and social imagery. Routledge, London

Breakwell GM (2007) The psychology of risk. Cambridge University Press, Cambridge

Calhoun C, Gerteis J, Moody J, Pfaff S, Virk I (2007) Contemporary sociological theory, 2nd edn. Blackwell, Oxford

Callon M, Lascoumes P, Barthe Y (2009) Acting in an uncertain world: an essay on technical democracy. MIT Press, Cambridge

Dean M (1999) Risk, calculable and incalculable. In: Lupton D (ed) Risk and sociocultural theory. Cambridge University Press, Cambridge, pp 131–159

Douglas M (1966) Purity and danger: an analysis of the concepts of pollution and taboo. Routledge Kegan Paul, London

Douglas M (1978) Cultural bias. Occasional Paper no. 35, Royal anthropological institute of Great Britain and Ireland, London

Douglas M (1982) Introduction to grid/group analysis. In: Douglas M (ed) Essays in the sociology of perception. Routledge Kegan Paul, London, pp 1–8

Douglas M (1992) Risk and blame: essays in cultural theory. Routledge, London

Douglas M (1996) Thought styles: critical essays on good taste. Sage, London

Douglas M, Wildavsky A (1982) Risk and culture: an essay on the selection of technological and environmental dangers. University of California Press, Berkeley

Durkheim E (1982) The rules of sociological method. Macmillan Press, New York

Eriksson B (1993) The first formulation of sociology: a discursive innovation of the 18th century. Eur J Sociol 34(2):251–276

Ferguson A (1767/1996) Essay on the history of civil society. Cambridge University Press, New York

Finucane M, Slovic P, Mertz CK, Flynn J, Satterfield T (2000) Gender, race and perceived risk: the "white male" effect. Health Risk Soc 2(2):159–172

Foucault M (1991) Governmentality. In: Burchell G, Gordon C, Miller P (eds) The Foucault effect: studies in governmentality. Harvester Wheatsheaf, Hemel Hempstead, pp 87–104

Funtowicz SO, Ravetz JR (1990) Uncertainty and quality in science for policy. Kluwer, Dordrecht

Furedi F (2002) The culture of fear risk-taking and the morality of low expectation. Continuum, London

Furedi F (2008) Fear and security: a vulnerability-led policy response. Soc Policy Adm 42(6):645–661

Giddens A (1984) The constitution of society: outline of the theory of structuration. Polity, Cambridge

Giddens A (1990) The consequences of modernity. Polity, Cambridge

Giddens A (1991) Modernity and self-identity. Polity, Cambridge

Giddens A (1994) Risk, trust and reflexivity. In: Beck U, Giddens A, Lash S (eds) Reflexive modernization. Politics, tradition and aesthetics in the modern social order. Polity, Cambridge, pp 184–197

Giddens A (1999) Runaway world: how globalization is reshaping our lives. Routledge, New York

Gieryn TF (1999) Cultural boundaries of science: credibility on the line. University of Chicago Press, Chicago

Gouldson A, Lidskog R, Wester-Herber M (2007) The battle for hearts and minds: evolutions in organisational approaches to environmental risk communication. Environ Plann C 25(1):56–72

Gutteling J, Wiegman O (1996) Exploring risk communication. Kluwer, Dordrecht

Hilgartner S (1992) The social construction of risk objects: or, how to pry open networks of risks. In: Short JF, Clarke L (eds) Organizations, uncertainties, and risk. Westview Press, Boulder, pp 39–53

Hilgartner S (2000) Science on stage: expert advice as public drama. Stanford University Press, Stanford

Hobbes T (1651/2005) Leviathan. Continuum, London

Höijer B, Lidskog R, Uggla Y (2006) Facing dilemmas. Sense-making and decision-making in late moder- nity. Futures 38(3):350–366

Hood C, Rothstein H, Baldwin R (2001) The government of risk: understanding risk regulation regimes. Oxford University Press, Oxford

Howel D, Moffatt S, Prince H, Bush J, Dunn C (2002) Urban air quality in North-East England: exploring the influences on local views and perceptions. Risk Anal 22(1):121–130

Hughes E, Kitzinger J, Murdock G (2006) The media and risk. In: Taylor-Gooby P, Zinn J (eds) Risk in social sciences. Oxford University Press, Oxford, pp 250–270

Hutter BM (2001) Regulation and risk. Occupational health and safety on the railways. Oxford University Press, Oxford

Hutter B, Power M (eds) (2005) Organizational encounters with risk. Cambridge University Press, Cambridge

Irwin A (2006) The politics of talk: coming to terms with the "new" scientific governance. Soc Stud Sci 36(2):299–320

Irwin A, Michael M (2003) Science, social theory and public knowledge. Open University Press, Maidenhead

Irwin A, Wynne B (eds) (1996) Misunderstanding science? The public reconstruction of science and technology. Cambridge University Press, Cambridge

Jaeger CC, Renn O, Rosa EA, Webler T (2001) Risk, uncertainty and rational action. Earthscan, London

Japp KP, Kusche I (2008) System theory and risk. In: Zinn J (ed) Social theories of risk and uncertainty: an introduction. Blackwell, Oxford, pp 76–105

Jasanoff S (ed) (2004) States of knowledge: the co-production of science and social order. Routledge, London

Jasanoff S (2005) Designs of nature: science and democracy in Europe and the United States. Princeton University Press, Princeton

Jasanoff S (2010) A new climate for society. Theory Culture Soc 27(2–3):233–253

Kasperson RE (1992) The social amplification of risk: progress in developing an integrative framework of risk. In: Krimsky S, Golding D (eds) Social theories of risk. Praeger, Westport

Kasperson RE, Renn O, Slovic P, Brown HS, Emel J, Goble R, Kasperson JX, Ratick S (1988) The social amplification of risk: a conceptual framework. Risk Anal 8(2):177–187

Kasperson JX, Kasperson RE, Pidgeon N, Slovic P (2003) The social amplification of risk: assessing fifteen years of research and theory. In: Pidgeon N, Kasperson RE, Slovic P (eds) Social amplification of risk. Cambridge University Press, Cambridge, pp 13–46

Kemshall H (2006) Crime and risk. In: Taylor-Gooby P, Zinn J (eds) Risk in social sciences. Oxford University Press, Oxford, pp 76–93

Krimsky S, Golding D (eds) (1992) Social theories of risk. Praeger, Westport

Latour B (1993) We have never been modern. Harvester Wheatsheaf, New York

Latour B (2004) Politics of nature: how to bring the sciences into democracy. Harvard University Press, Cambridge

Latour B (2005) Reassembling the social: an introduction to actor-network theory. Oxford University Press, Oxford

Levinson R, Thomas J (eds) (1997) Science today: problem or crisis?. Routledge, London

Lidskog R (1996) In science we trust? On the relation between scientific knowledge, risk consciousness and public trust. Acta Sociologica 39(1):31–56

Lidskog R (2008) Scientised citizens and democratised science. Re-assessing the expert-lay divide. J Risk Res 11(1–2):69–86

Lidskog R (2011) Regulating nature: public understanding and moral reasoning. Nat Cult 6(2):149–167

Lidskog R, Sundqvist G (2011) The science-policy-citizen dynamics in international environmental governance. In: Lidskog R, Sundqvist G (eds) Governing the air: science-policy-citizen dynamics in international environmental governance. MIT Press, Cambridge, pp 323–359

Lidskog R, Soneryd L, Uggla Y (2005) Knowledge, power and control: studying environmental regulation in late modernity. J Environ Policy Plann 7(2):89–106

Lidskog R, Soneryd L, Uggla Y (2009) Transboundary risk governance. Earthscan, London

Lidskog R, Uggla Y, Soneryd L (2011) Making transboundary risks governable: reducing complexity, constructing identities and ascribing capabilities. Ambio 40(2):111–120

Lofstedt RE (2005) Risk management in post-trust societies. Palgrave Macmillan, New York

Luhmann N (1984) Soziale systeme. Grundriss einer allgemeinen theorie. Suhrkamp, Frankfurt a M

Luhmann N (1989) Ecological communication. Polity, Cambridge

Luhmann N (1993) Risk: a sociological theory. Walter de Gruyter, Berlin

Lupton D (1999) Introduction: risk and sociocultural theory. In: Lupton D (ed) Risk and sociocultural theory. Cambrdge University Press, Cambridge, pp 1–11

Marres N (2007) The issues deserve more credit: pragmatist contributions to the study of public involvement in controversy. Soc Stud Sci 37(5):759–778

Millar J (1771) Observations concerning the distinctions of rank in society. T Ewing, Edinburgh

Novas C, Rose N (2000) Genetic risk and the birth of the somatic individual. Econ Soc 29(4):485–513

O'Malley P (2008) Risk and governmentality. In: Zinn JO (ed) Social theories of risk and uncertainty: an introduction. Blackwell, Oxford, pp 52–75

Pidgeon N, Kasperson RE, Slovic P (eds) (2003) Social amplification of risk. Cambridge University Press, Cambridge

Power M (2007) Organized uncertainty: designing a world of risk management. Oxford University Press, Oxford

Reith G (2009) Uncertain times: the notion of "risk" and the development of modernity. In: Löfstedt RE, Boholm A (eds) The Earthscan reader on risk. Earthscan, London, pp 53–68

Renn O (1998) Three decades of risk research: accomplishments and new challenges. J Risk Res 1(1):49–71

Renn O (2008) Risk governance: coping with uncertainty in a complex world. Earthscan, London

Rosa EA (1998) Metatheoretical foundations for post-normal risk. J Risk Res 1(1):15–44

Rose N (1993) Government, authority and expertise in advanced liberalism. Econ Soc 22:283–299

Rose N (1996) Governing "advanced" liberal democracies. In: Barry A, Osborne T, Rose N (eds) Foucault and political reason. UCL Press, London, pp 37–64

Rose N (1999) Powers of freedom: reframing political thought. Cambridge University Press, Cambridge

Rose N, Miller P (1992) Political power beyond the states: problematics of government. Br J Sociol 43(2):173–205

Sjöberg L (2000) Perceived risk and tampering with nature. J Risk Res 3(4):353–367

Slovic P (1987) Perceptions of risk. Science 236:280–285

Slovic P, Peters E (1998) The importance of worldviews in risk perception. Risk Decis Policy 3(2):165–170

Smith A (1776/2001) An inquiry into the nature and causes of the wealth of nation. Random House International, New York

Starr C (1969) Social benefit versus technological risk. Science 165(3899):1232–1238

Stehr N (2005) Knowledge politics: governing the consequences of science and technology. Paradigm, Boulder

Sundqvist G, Elam M (2010) Public involvement designed to circumvent public concern? The 'participatory turn' in European nuclear activities (Article 8). Risk Hazards Crisis Public Policy 1(4):203–229

Taylor-Gooby P, Zinn J (eds) (2006) Risk in social sciences. Oxford University Press, Oxford

Thompson M, Ellis R, Wildavsky A (1990) Cultural theory. Westview Press, Boulder

Todt O (2003) Designing trust. Futures 35:239–251

Tulloch J (1999) Fear of crime and the media: sociocultural theories of risk. In: Lupton D (ed) Risk and sociocultural theory. Cambridge University Press, Cambridge, pp 34–58

Tulloch J, Lupton D (2003) Risk and everyday life. Sage, London

Wester-Herber M (2004) Underlying concerns in land-use conflict: the role of place-identity in risk perception. Environ Sci Policy 7(2):109–116

Wynne B (1992a) Misunderstood misunderstanding: social identities and public uptake of science. Public Underst Sci 1(3):281–304

Wynne B (1992b) Uncertainty and environmental learning: reconceiving science and policy in the preventive paradigm. Glob Environ Chang 2(2):111–127

Wynne B (1993) Public uptake of science: a case for institutional reflexivity. Public Underst Sci 2(4):321–337

Wynne B (1995) Public understanding of science. In: Jasanoff S, Markle GE, Peterson JC, Pinch T (eds) Handbook of science and technology studies. Sage, Thousand Oaks, pp 361–388

Wynne B (1996) May the sheep safely graze? A reflexive view of the expert-lay knowledge divide. In: Lash S, Szerszynsky B, Wynne B (eds) Risk, environment modernity: towards a new ecology. Sage, London, pp 44–83

Wynne B (2001) Creating public alienation: expert cultures of risk and ethics on GMOs. Sci Cult 10(4):445–481

Wynne B (2005) Risk as globalizing 'democratic' discourse? Framing subjects and citizens. In: Leach M, Scoones I, Wynne B (eds) Science and citizens: globalization and the challenge of engagement. ZED Books, London, pp 66–82

Wynne B (2010) Strange weather, again: climate science as political art. Theor Cult Soc 27(2–3):289–305

Zinn JO (2008) A comparison of sociological theorizing on risk and uncertainty. In: Zinn JO (ed) Social theories of risk and uncertainty: an introduction. Blackwell, Oxford, pp 168–210

Zinn JO, Gooby-Taylor P (2006) Risk as an interdisciplinary research area. In: Taylor-Gooby P, Zinn J (eds) Risk in social science. Oxford University Press, Oxford

Chapter 5
Risk and Responsibility

Ibo van de Poel and Jessica Nihlén Fahlquist

Abstract When a risk materializes, it is common to ask the question: who is responsible for the risk being taken? Despite this intimate connection between risk and responsibility, remarkably little has been written on the exact relation between the notions of risk and responsibility. This contribution sets out to explore the relation between risk and responsibility on basis of the somewhat dispersed literature on the topic and it sketches directions for future research. It deals with three more specific topics. First we explore the conceptual connections between risk and responsibility by discussing different conceptions of risk and responsibility and their relationships. Second, we discuss responsibility for risk, paying attention to four more specific activities with respect to risks: risk reduction, risk assessment, risk management, and risk communication. Finally, we explore the problem of many hands (PMH), that is, the problem of attributing responsibility when large numbers of people are involved in an activity. We argue that the PMH has especially become prominent today due to the increased collective nature of actions and due to the fact that our actions often do not involve direct harm but rather risks, that is, the possibility of harm. We illustrate the PMH for climate change and discuss three possible ways of dealing with it: (1) responsibility-as-virtue, (2) a procedure for distributing responsibility, and (3) institutional design.

I. van de Poel (✉) · J. Nihlén Fahlquist
Delft University of Technology, Delft, The Netherlands
e-mail: i.r.vandepoel@tudelft.nl

J. N. Fahlquist
Royal Institute of Technology, Stockholm, Sweden
e-mail: j.a.nihlen-fahlquist@tudelft.nl

S. Roeser et al. (eds.), *Essentials of Risk Theory*, SpringerBriefs in Philosophy,
DOI: 10.1007/978-94-007-5455-3_5, © The Author(s) 2013

Introduction

Risk and responsibility are central notions in today's society. When the Deepwater Horizon oil rig exploded in April 2010 killing 11 people and causing a major oil spill in the Gulf of Mexico, questions were asked whether no unacceptable risks had been taken and who was responsible. The popular image in cases like this appears to be that if such severe consequences occur, someone must have, deliberately or not, taken an unacceptable risk and for that reason that person is also responsible for the outcome. One reason why the materialization of risks immediately raises questions about responsibility is our increased control over the environment. Even in cases of what are called natural risks, that is, risks with primarily natural rather than human causes, questions about responsibility seem often appropriate nowadays. When an earthquake strikes a densely populated area and kills thousands of people, it may be improper to hold someone responsible for the mere fact that the earthquake occurred, but it might well be appropriate to hold certain people responsible for the fact that no proper warning system for earthquakes was in place or for the fact that the buildings were not or insufficiently earthquake resistant. In as far as both factors mentioned contributed to the magnitude of the disaster, it might even be appropriate to hold certain people responsible for the fatalities.

The earthquake example shows that the idea that it is by definition impossible to attribute responsibility for natural risks and that such risks are morally less unacceptable is increasingly hard to maintain, especially due to technological developments. This may be considered a positive development in as far it has enabled mankind to drastically reduce the number of fatalities, and other negative consequences, as a result of natural risks. At the same time, technological development and the increasing complexity of society have introduced new risks; the Deepwater Horizon oil rig is just one example. Especially in the industrialized countries, these new risks now seem to be a greater worry than the traditional so-called natural risks. Although these new risks are clearly man-made, they are in practice not always easy to control. It is also often quite difficult to attribute responsibility for them due to the larger number of people involved; this is sometimes referred to as the "problem of many hands" (PMH), which we will describe and analyze in more detail in section Further Research: Organizing Responsibility for Risks. Before we do so, we will first discuss the relation between risk and responsibility on a more abstract, conceptual level by discussing different conceptions of risk and responsibility and their relation, in section Conceptions of Risk and Responsibility. Section Responsibility for Risks focuses on the responsibility for dealing with risk; it primarily focuses on so-called forward-looking moral responsibility and on technological risks.

While risk and responsibility are central notions in today's society and a lot has been written about both, remarkably few authors have explicitly discussed the relation between the two. Moreover, the available literature is somewhat dispersed over various disciplines, like philosophy, sociology, and psychology. As a consequence, it is impossible to make a neat distinction between the established state of the art and future research in this contribution. Rather the contribution as a whole has a somewhat explorative character. Nevertheless, sections Conceptions of Risk and Responsibility and Responsibility for Risks mainly discuss the existing literature, although they make some connections that cannot be found in the current literature. Section Further Research: Organizing Responsibility for Risks explores the so-called problem of many hands and the need to organize responsibility, which is rather recent and requires future research, although some work has already been done and possible directions for future research can be indicated.

Conceptions of Risk and Responsibility

Both risk and responsibility are complex concepts that are used in a multiplicity of meanings or conceptions as we will call them. Moreover, as we will see below, while some of these conceptions are merely descriptive, others are clearly normative. Before we delve deeper into the relation between risk and responsibility, it is therefore useful to be more precise about both concepts. We will do so by first discussing different conceptions of risk (section Conceptions of Risk) and of responsibility (section Conceptions of Responsibility). We use the term "conception" here to refer to the specific way a certain concept like risk or responsibility is understood. The idea is that while different authors, approaches, or theories may roughly refer to the same concept, the way they understand the concept and the conceptual relations they construe with other concepts is different. After discussing some of the conceptions of risk and responsibility, section Conceptual Relations Between Risk and Responsibility discusses conceptual relations between risk and responsibility.

Conceptions of Risk

The concept of risk is used in different ways (see "The Concepts of Risk and Safety"). Hansson (2009, pp. 1069–1071), for example, mentions the following conceptions:

1. Risk = an *unwanted event* that may or may not occur
2. Risk = the *cause* of an unwanted event that may or may not occur
3. Risk = the *probability* of an unwanted event that may or may not occur
4. Risk = the statistical expectation value of an unwanted event that may or may not occur
5. Risk = the fact that a decision is made under conditions of *known probabilities* ("decision under risk")

The fourth conception has by now become the most common technical conception of risk and this conception is used usually in engineering and in risk assessment. The fifth conception is common in decision theory. In this field, it is common to distinguish decisions under risk from decisions under certainty and decisions under uncertainty. Certainty refers to the situation in which the outcomes (or consequences) of possible actions are certain. Risk refers to the situation in which possible outcomes are known and the probabilities (between 0 and 1) of occurrence of these outcomes are known. Uncertainty refers to situations in which possible outcomes are known but no probabilities can be attached to these outcomes. A situation in which even possible outcomes are unknown may be referred to as ignorance.

The fifth, decision-theoretical conception of risk is congruent with the fourth conception in the sense that both require knowledge of possible outcomes and of the probability of such outcomes to speak meaningfully about a risk. One difference is that whereas the fifth conception does not distinguish between wanted and unwanted outcomes, the fourth explicitly refers to unwanted outcomes. Both the fourth and the fifth conception are different from the way the term "risk" is often used in daily language. In daily language, we commonly refer to an undesirable event as a risk, even if the probability is unknown or the exact consequences are unknown. One way to deal with this ambiguity is to distinguish between hazards (or dangers) and risks. Hazard refers to the mere possibility of an unwanted event (conception 1 above), without necessarily knowing either the consequences or the probability of such an unwanted event. Risk may then be seen as a specification of the notion of hazard. The most common definition of risk in engineering and risk assessment, and more generally in techno-scientific contexts, is that of statistical expectation value, or the product of the consequences of an unwanted event and the probability of the unwanted event occurring (meaning 4 above). But even in techno-scientific contexts other definitions of risk can be found. The International Program on Chemical Safety, for example, in an attempt to harmonize the different meanings of terms used in risk assessment defines risk as: "The probability of an adverse effect in an organism, system, or (sub)population caused under specified circumstances by exposure to an agent" (International Program on Chemical Safety 2004, p. 13). This is closer to the third than the fourth conception mentioned by Hansson. Nevertheless, the International Program on Chemical Safety appears to see risk as a further specification of hazard, which they define as: "Inherent property of an agent or situation having the potential to cause adverse effects when an organism, system, or (sub)population is exposed to that agent" (International Program on Chemical Safety 2004, p. 12).

Conceptions of risk cannot only be found in techno-scientific contexts and in decision theory, but also in social science, in literature on risk perception (psychology), and more recently in moral theory (for a discussion of different conceptions of risk in different academic fields, see (Bradbury 1989; Thompson and Dean 1996; Renn 1992; Shrader-Frechette 1991a). We will below discuss some of the main conceptions of risk found in these bodies of literature. The technical conception of risk assumes, at least implicitly, that the only relevant aspects of risk are the magnitude of certain unwanted consequences and the probability of these consequences occurring. The conception nevertheless contains a normative element because it refers to *unwanted* consequences (or events). However, apart from this normative element, the conception is meant to be descriptive rather than normative. Moreover, it is intended to be context free, in the sense that it assumes that the only relevant information about a risky activity is the probability and magnitude of consequences (Thompson and Dean 1996). Typically, conceptions of risk in psychology, social science, and moral theory are more contextual. They may refer to such contextual information as by whom the risk is run, whether the risk is imposed or voluntary, whether it is a natural or man-made risk, and so on. What contextual elements are included, and the reason for which contextual elements are included is, however, different for different contextual conceptions of risk.

The psychological literature on risk perception has established that lay people include contextual elements in how they perceive and understand risks (e.g., Slovic 2000). These include, for example, dread, familiarity, exposure, controllability, catastrophic potential, perceived benefits, time delay (future generations), and voluntariness. Sometimes the fact that lay people have a different notion of risk than experts, and therefore estimate the magnitude of risks differently, is seen as a sign of their irrationality. This interpretation assumes that the technical conception of risk is the right one and that lay people should be educated to comply with it. Several authors have, however, pointed out that the contextual elements included by lay people are relevant for the acceptability of risks and for risk management and that in that sense the public's conception of risk is "richer" and in a sense more adequate than that of scientific experts (e.g., Slovic 2000; Roeser 2006, 2007). In the literature on the ethics of risk it is now commonly accepted that the moral acceptability of risks depends on more concerns than just the probability and magnitude of possible negative consequences (see "Ethics and Risk"). Moral concerns that are often mentioned include voluntariness, the balance and distribution of benefits and risks (over different groups and over generations), and the availability of alternatives (Asveld and Roeser 2009; Shrader-Frechette 1991b; Hansson 2009; Harris et al. 2008; Van de Poel and Royakkers 2011).

In the social sciences, a rich variety of conceptions of risk have been proposed (Renn 1992). We will not try to discuss or classify all these conceptions, but will briefly outline two influential social theories of risk, that is, cultural theory (Douglas and Wildavsky 1982) and risk society (Beck 1992). Cultural theory conceives of risks as collective, cultural constructs (see "Cultural Cognition as a Conception of the Cultural Theory of Risk"). Douglas and Wildavsky (1982) distinguish three cultural biases that correspond to and are maintained by three types of social organization: hierarchical, market individualistic, and sectarian. They claim that each bias corresponds to a particular selection of dangers as risks. Danger here refers to what we above called a hazard: the (objective) possibility of something going wrong. According to Douglas and Wildavsky, dangers cannot be known directly. Instead they are culturally constructed as risks. Depending on the cultural bias, certain dangers are preeminently focused on. Hierarchists focus on risks of human violence (war, terrorism, and crime), market individualists on risks of economic collapse, and sectarians on risks of technology (Douglas and Wildavsky 1982, pp. 187–188).

Like Douglas and Wildavsky, Ulrich Beck in his theory of risk society sees risk as a social construct. But whereas Douglas and Wildavsky focus on the cultural construction of risks and believe that various constructions may exist side by side, Beck places the social construction of risk in historical perspective. Beck defines risk as "a systematic way of dealing with hazards and insecurities induced and introduced by modernization itself" (Beck 1992, p. 21, emphasis in the original). Speaking in terms of risks, Beck claims, is historically a recent phenomenon and it is closely tied to the idea that risks depend on decisions (Beck 1992, p. 183). Typically for what Beck calls the "risk society" is that it has become impossible to attribute hazards to external causes. Rather, all hazards are seen as depending on human choice and, hence, are, according to Beck's definition of these notions, conceived as risks. Consequently, in risk society the central issue is the allocation of risk rather than the allocation of wealth as it was in industrial society.

Some authors have explicitly proposed to extend the technical conception of risk to include some of the mentioned contextual elements. We will briefly outline two examples. Rayner (1992) has proposed the following adaption to the conventional conception of risk:

$$R = (P \times M) + (T \times L \times C)$$

with

R   Risk
P   Probability of occurrence of the adverse event
M   Magnitude of the adverse consequences
T   Trustworthiness of the institutions regulating the technology
L   Acceptability of the principle used to apportion liabilities for undesired
     consequences
C   Acceptability of the procedure by which collective consent is obtained to
     those who must run the consequences

Although this conception has a number of technical difficulties, it brings to the fore some of the additional dimensions that are important not just for the perception or cultural construction of risks but also for their regulation and moral acceptability

More recently, Wolff (2006) has proposed to add cause as a primary variable in addition to probability and magnitude to the conception of risk. The rationale for this proposal is that cause is also relevant for the acceptability of risks. Not only may there be a difference between natural and man-made risks, but also different man-made risks may be different in acceptability depending on whether the human cause is based on culpable or non-culpable behavior and the type of culpable behavior (e.g., malice, recklessness, negligence, or incompetence). We might have good moral reasons to consider risks based on malice (e.g., a terrorist attack) less acceptable than risks based on incompetence even when they are roughly the same in terms of probability and consequences. In addition to cause, Wolff proposes to add such factors as fear (dread), blame, and shame as secondary variables that might affect each of the primary variables. Like in the case of Rayner's conception, the technicalities of the new conception are somewhat unclear, but it is definitively an attempt to broaden the conception of risk to include contextual elements that are important for the (moral) acceptability of risks.

Rayner's and Wolff's proposals raise the question whether all factors which are relevant for decisions about acceptable risk or risk management should be included in a conceptualization of risk. Even if it is reasonable to include moral concerns in our decisions about risks, it may be doubted whether the best way to deal with such additional concerns is to build them into a (formal) conception of risk.

Conceptions of Responsibility

Like the notion of risk, the concept of responsibility can be conceptualized in different ways. One of the first authors to distinguish different conceptions of responsibility was Hart (1968, pp. 210–237) who mentions four main conceptions of responsibility: role responsibility, causal responsibility, liability responsibility, and capacity responsibility. Later authors have distinguished additional conceptions, and the following gives a good impression of the various conceptions that might be distinguished (Van de Poel 2011):

1. Responsibility-as-cause. As in: the earthquake caused the death of 100 people.
2. Responsibility-as-role. As in: the train driver is responsible for driving the train.
3. Responsibility-as-authority. As in: he is responsible for the project, meaning he is in charge of the project. This may also be called responsibility-as-office or responsibility-as-jurisdiction. It refers to a realm in which one has the authority to make decisions or is in charge and for which one can be held accountable.
4. Responsibility-as-capacity. As in: the ability to act in a responsible way. This includes, for example, the ability to reflect on the consequences of one's actions, to form intentions, to deliberately choose an action and act upon it.
5. Responsibility-as-virtue, as the disposition (character trait) to act responsibly. As in: he is a responsible person. (The difference between responsibility-as-capacity and responsibility-as-virtue is that whereas the former only refers to ability, the second refers to a disposition that is also surfacing in actions. So someone who has the capacity for responsibility may be an irresponsible person in the virtue sense).
6. Responsibility-as-obligation to see to it that something is the case. As in: he is responsible for the safety of the passengers, meaning he is responsible to see to it that the passengers are transported safely.
7. Responsibility-as-accountability. As in: the (moral obligation) to account for what you did or what happened (and your role in it happening).

8. Responsibility-as-blameworthiness. As in: he is responsible for the car
   accident, meaning he can be properly blamed for the car accident happening.
9. Responsibility-as-liability. As in: he is liable to pay damages.

The first four conceptions are more or less descriptive: responsibility-as-cause,
role, authority, or capacity describes something that is the case or not. The other
five are more normative. The first two normative conceptions—responsibility-as-
virtue and responsibility-as-obligation—are primarily forward-looking (prospective) in nature. Responsibility-as-accountability, blameworthiness, and liability are
backward-looking (retrospective) in the sense that they usually apply to something
that has occurred. Both the forward-looking and the backward-looking normative
conception of responsibility are relevant in relation to risks. Backward-looking
responsibility is mainly at stake when a risk has materialized and then relates to
such questions like: Who is accountable for the occurrence of the risk? Who can be
properly blamed for the risk? Who is liable to pay the damage resulting from the
risk materializing? Forward-looking responsibility is mainly relevant with respect
to the prevention and management of risks. It may refer to different tasks that are
relevant for preventing and managing risk like risk assessment, risk reduction, risk
management, and risk communication. We will discuss the responsibility for these
tasks in section Responsibility for Risks.

Cross-cutting the distinction between the different conceptions of responsibility
is a distinction between what might be called different "sorts" of responsibility like
organizational responsibility, legal responsibility, and moral responsibility. The
main distinction between these sorts is the grounds on which it is determined
whether someone is responsible (in one of the senses distinguished above).
Organizational responsibility is mainly determined by the rules and roles that exist
in an organization, legal responsibility by the law (including jurisprudence), and
moral responsibility is based on moral considerations. The two types of distinctions
are, however, not completely independent of each other. Organizational
responsibility, for example, often refers to responsibility-as-task or responsibility-
as-authority and seems unrelated to responsibility-as-cause and responsibility-
as-capacity. It might also refer to responsibility-as-accountability, just like legal
and moral responsibility. We might thus distinguish between organizational, legal,
and moral accountability, where the first is dependent on an organization's rules
and roles, the second on the law, and the third on moral considerations.

In this contribution we mainly focus on moral responsibility. Most of the
general philosophical literature on responsibility has focused on backward-looking
moral responsibility, in particular on blameworthiness. In this literature also, a
number of general conditions have been articulated which should be met in order
for someone to be held properly or fairly responsible (e.g., Wallace 1994; Fischer
and Ravizza 1998). Some of these conditions, especially the freedom and
knowledge condition, go back to Aristotle (*The Nicomachean Ethics*, book III,
Chaps. 1–5). These conditions include:

1. Moral agency. The agent A is a moral agent, that is, has the capacity to act
   responsibly (responsibility-as-capacity).

2. Causality. The agent A is somehow causally involved in the action or outcome for which A is held responsible (responsibility-as-cause).
3. Wrongdoing. The agent A did something wrong.
4. Freedom. The agent A was not compelled to act in a certain way or to bring about a certain outcome.
5. Knowledge. The agent A knew, or at least could reasonably have known that a certain action would occur or a certain outcome would result and that this was undesirable.

Although these general conditions can be found in many accounts, there is much debate about at least two issues. One is the exact content and formulation of each of the conditions. For example, does the freedom condition imply that the agent could have acted otherwise (e.g., Frankfurt 1969)? The other is whether these conditions are individually necessary and jointly sufficient in order for an agent to be blameworthy. One way to deal with the latter issue is to conceive of the mentioned conditions as arguments or reasons for holding someone responsible (blameworthy) for something rather than as a strict set of conditions (Davis 2012).

Whereas the general philosophical literature on responsibility has typically focused on backward-looking responsibility, the more specific analyses of moral responsibility in technoscientific contexts, and more specifically as applied to (technological) risks, often focus on forward-looking responsibility. They, for example, discuss the forward-looking responsibility of engineers for preventing or reducing risks (e.g., Davis 1998; Harris et al. 2008; Martin and Schinzinger 2005; Van de Poel and Royakkers 2011). One explanation for this focus may be that in these contexts the main aim is to prevent and manage risks rather than to attribute blame and liability. This is, of course, not to deny that in other contexts, backward-looking responsibility for risks is very relevant. It surfaces, for example, in court cases about who is (legally) liable for certain damage resulting from the materialization of technological risks. It is also very relevant in more general social and political discussions about how the costs of risks should be borne: by the victim, by the one creating the risks, or collectively by society, for example, through social insurance.

Conceptual Relations Between Risk and Responsibility

The conceptual connections between risk and responsibility depend on which conception of risk and which conception of responsibility one adopts. The technical conception of risk, which understands risks as the product of probability and magnitude of certain undesirable consequences, is largely descriptive, but it contains a normative element because it refers to undesirable outcomes. Typically, responsibility also is often used in reference to undesirable outcomes, especially if responsibility is understood as blameworthiness. Yet if the undesirable consequences, to which the technical conception of risk refers, materialize this does not necessarily imply that someone is blameworthy for these consequences. As we have seen, a number of conditions have to be met in order for someone to fairly be held responsible for such consequences. In cases of risks the knowledge condition will usually be fulfilled because if a risk has been established it is known that certain consequences might occur. It will often be less clear whether the wrongdoing condition is met. Risks normally refer to unintended, but not necessarily unforeseen, consequences of action. Nevertheless, under at least two circumstances, the introduction of a risk amounts to wrongdoing. One is if the actor is reckless, that is, if he knows that a risk is (morally) unacceptable but still exposes others to it. The other is negligence. In the latter case, the actor is unaware of the risk he is taking but should and could have known the risk and exposing others to the risk is unacceptable.

If we focus on forward-looking responsibility rather than backward-looking responsibility, the technical conception of risk might be thought to imply an obligation to avoid risks since most conceptions of risk refer to something undesirable. Again, however, the relation is not straightforward. Some risks, like certain natural risks, may be unavoidable. Other risks may not be unavoidable but worth taking given the advantages of certain risky activities. Nevertheless, there seems to be a forward-looking responsibility to properly deal with risks. In section Responsibility for Risks, we will further break down that responsibility and discuss some of its main components.

In the psychological literature on risk perception, no direct link is made between risk and responsibility. Nevertheless, some of the factors that this literature has shown to influence the perception of risk may be linked to the concept of responsibility. One such factor is controllability (e.g., Slovic 2000). Control is often seen as a precondition for responsibility; it is linked to the conditions of freedom and knowledge we mentioned above. Also voluntariness, another important factor in the perception of risk (e.g., Slovic 2000), is linked to those responsibility conditions. This suggests that risks for which one is not responsible (or cannot take responsibility) but to which one is exposed beyond one's will and/or control are perceived as larger and less acceptable.

In the sociological literature on risk that we discussed in section Conceptions of Risk, a much more direct connection between risk and responsibility is supposed. Mary Douglas (1985) argues that the same institutionally embedded cultural biases that shape the social construction of risks also shape the attribution of responsibility, especially of blameworthiness. Institutions are, according to Douglas, typically characterized by certain recurring patterns of attributing blame, like blaming the victim, or blaming outsiders, or just accepting the materialization of risks as fate or the price to be paid for progress. According to the theory of risk society, both risk and responsibility are connected to control and decisions. This implies a rather direct conceptual connection between risk and responsibility. As Anthony Giddens expresses it:

> The relation between risk and responsibility can be easily stated, at least on an abstract level. Risks only exist when there are decisions to be taken.... The idea of responsibility also presumes decisions. What brings into play the notion of responsibility is that someone takes a decision having discernable consequences (Giddens 1999, p. 8).

The sociological literature seems to refer primarily to organizational responsibility, in the sense that attribution of responsibility primarily depends on social conventions. Nevertheless, as we have seen the idea of control, which is central to risk in the theory of risk society, is also central for moral responsibility.

The redefinitions of risk proposed by Rayner and Wolff, finally, both refer to responsibility as an ingredient in the conception of risk. Rayner includes liability as an aspect of risk. While liability is usually primarily understood as a legal notion, his reference to the *acceptability* of liability procedures also has clear moral connotations. In Wolff's conception of risk, responsibility affects the variable "cause" that he proposes as additional primary variable for risk. As Wolff points out, it matters for the acceptability of risk whether it is caused by malice, recklessness, or negligence. These distinctions also have a direct bearing on the moral responsibility of the agent causing the undesirable consequences; they represent different degrees of wrongdoing. So, on Wolff's conceptualization, whether and to what degree anyone is responsible for a risk has a bearing on the acceptability of that risk.

Although the relation between risk and responsibility depends on the exact conceptualization of both terms and one might discuss how to best conceptualize both terms, the above discussion leads to a number of general conclusions. First, if an undesired outcome is the result of someone taking a risk or exposing others to a risk, it appears natural to talk about responsibility in the backward-looking sense (accountability, blameworthiness, liability) for those consequences and for the risk taken. Second, both risk and responsibility are connected to control and decisions. Even if one does not accept the tight conceptual connection between risk and control that the theory of risk society supposes, it seems clear that risks often are related to decisions and control. As pointed out in the introduction, even so-called natural risks increasingly come under human control. This implies that we cannot only hold people responsible for risks in a backward-looking way, but that people can also take or assume forward-looking responsibility (responsibility-as-virtue or as obligation) for risks. Third, the acceptability of risks appears to depend, at least partly, on whether someone can fairly be held responsible for the risk occurring or materializing.

Responsibility for Risks

In the literature on risk some general frameworks have been developed for thinking about the responsibility for risks and some general tentative answers have been formulated to the question who is responsible for certain risks. In this section, we present a number of these positions and the debates to which they have given rise. We focus on human-induced risks, that is, nonnatural risks, with a prime focus on technological risks. Our focus is also primarily on forward-looking responsibility rather than on backward-looking responsibility (accountability, blameworthiness, and liability) for risks.

Forward-looking responsibility for risks can be subdivided in the following main responsibilities:

1. Responsibility for risk reduction.
2. Responsibility for risk assessment, that is, establishing risks and their magnitude.
3. Responsibility for risk management. Risk management includes decisions about what risks are acceptable and the devising of regulations, procedures, and the like to ensure that risks remain within the limits of what is acceptable.
4. Responsibility for risk communication, that is, the communication of certain risks, in particular to the public.

Section The Responsibility of Engineers will discuss the responsibility for risk reduction. In the case of technological risks, this responsibility is often attributed to engineers. Section Risk Assessment Versus Risk Management will focus on the responsibility for risk assessment versus risk management. The former is often attributed to scientists, while governments and company managers are often held responsible for the latter. It will be examined whether this division of responsibilities is justified. Section Individual Versus Collective Responsibility for Risks will focus on an important issue with respect to risk management: whether decisions concerning acceptable risk are primarily the responsibility of individuals who take and potentially suffer the risk or whether it is a collective responsibility that should be dealt with through regulation by the government. Section Risk Communication will discuss some of the responsibilities of risk communicators and related dilemmas that have been discussed in the literature on risk communication.

The Responsibility of Engineers

Engineers play a key role in the development and design of new technologies. In this role they also influence the creation of technological risks. In the engineering ethics literature, it is commonly argued that engineers have a responsibility for safety (Davis 1998; Harris et al. 2008; Martin and Schinzinger 2005; Van de Poel and Royakkers 2011). In this section, we will consider these arguments and discuss how safety and risk are related and what the engineers' responsibility for safety implies for their responsibility for technological risks.

Most engineering codes of ethics state that engineers have a responsibility for the safety of the public. Thus, the code of the National Society of Professional Engineers in the USA states that: "Engineers, in the fulfillment of their professional duties, shall…. Hold paramount the safety, health, and welfare of the public" (NSPE 2007). Safety is not only stressed as the engineer's responsibility in codes of ethics but also in technical codes and standards. Technical codes are legal requirements that are enforced by a governmental body to protect safety, health, and other relevant values (Hunter 1997). Technical standards are usually recommendations rather than legal requirements that are written by engineering experts in standardization committees. Standards are usually more detailed than technical codes and may contain detailed provisions about how to design for safety.

Does the fact that safety is a prime concern in engineering codes of ethics and technical codes and standards entail that engineers have a moral responsibility for safety? One can take different stances here. Some authors have argued that codes of ethics entail an implicit contract either between a profession and society or among professionals themselves. Michael Davis, for example, defines a profession as "a number of individuals in the same occupation voluntarily organized to earn a living by openly serving a certain moral ideal in a morally permissible way beyond what law, market, and morality would otherwise require" (Davis 1998, p. 417). This moral idea is laid down in codes of ethics and thus implies, as we have seen, a responsibility for safety. According to Davis, codes are binding because they are an implicit contract between professionals, to which engineers subscribe by joining the engineering profession.

One could also argue that codes of ethics or technical codes and standards as such do not entail responsibilities for engineers but that they *express* responsibilities that are grounded otherwise. In that case, the engineers' responsibility for safety may, for example, be grounded in one of the general ethical theories like consequentialism, deontology, or virtue ethics. But if we believe that engineers have a moral responsibility for safety, does this also entail a responsibility for risks? To answer this question, we need to look a bit deeper into the conceptual relation between safety and risk (see "The Concepts of Risk and Safety"). In engineering, safety has been understood in different ways. One understanding is that safety means absolute safety and, hence, implies the absence of risk. In most contexts, this understanding is not very useful (Hansson 2009, p. 1074). Absolute safety is usually impossible and even if it would be possible it would in most cases be undesirable because eliminating risks usually comes at a cost, not only in monetary terms but also in terms of other design criteria like sustainability or ease of use. It is therefore better to understand safety in terms of "acceptable risk". One might then say that a technological device is safe if its associated risks are acceptable. What is acceptable will depend on what is feasible and what is reasonable. The notion of reasonableness refers here to the fact that reducing risks comes at a cost and that hence not all risk reductions are desirable.

So conceived, engineers may be said to be responsible for reducing risks to an acceptable level. What is acceptable, however, requires a normative judgment. This raises the question whether the engineer's responsibility for reducing risks to an acceptable level includes the responsibility to make a normative judgment on which risks are acceptable and which ones are not or that it is limited to meeting an acceptable risk level that is set in another way, for example, by a governmental regulator. The answer to this question may well depend on whether the engineers are designing a well-established technology for which safety standards have been set that are generally and publicly recognized as legitimate or that they are designing a radically new technology, like nanotechnology, for which existing safety standards cannot be applied straightforwardly and of which the hazards and risks are more uncertain anyway (for this distinction, see Van de Poel and Van Gorp 2006). In the former case, engineers can rely on established safety standards. In the latter case, such standards are absent. Therefore in the second case engineers and scientists also have some responsibility for judging what risks are acceptable, although they are certainly not the only party that is or should be involved in such judgments.

Risk Assessment Versus Risk Management

In the previous section we have seen that a distinction needs to be made between responsibility for risk reduction and responsibility for decisions about acceptable risks. Engineers have a responsibility for risk reduction but not necessarily or at least to a lesser degree a responsibility for deciding about acceptable risk. In this section we will discuss a somewhat similar issue in the division of responsibility for risk, namely, the responsibility for establishing the magnitude of risks (risk assessment) and decisions about the acceptability and management of risks (risk management). Traditionally risk assessment is seen as a responsibility of scientists, and risk management as a responsibility of governments and (company) managers (National Research Council 1983) (see "Risk Management in Technocracy"). In this section, we will discuss whether this division of labor and responsibility is tenable or not. In particular, we will focus on the question whether adequate risk assessment can be completely value free, as is often supposed, or, as has been argued by a number of authors, that it needs to rely on at least some value judgments.

One reason why risk assessment cannot be entirely value free is that in order to do a risk assessment a decision needs to be made on what risks to focus. Since, on the conventional technical conception of risk (see section Conceptions of Risk), risks are by definition undesirable, classifying something as a risk already involves a value judgment. It might be argued, nevertheless, that decisions about what is undesirable are to be made by risk managers and that risk assessors, as scientists, should then investigate all potential risks. In practice, however, a risk assessment cannot investigate all possible risks; a selection will have to be made and selecting certain risks rather than others implies a value judgment. Again, it can be argued that this judgment is to be made by risk managers. A particular problem here might be that some risks are harder to investigate or establish scientifically than others. Some risks may even be statistically undetectable (Hansson 2009, pp. 1084–1086). From the fact that a risk is hard or even impossible to detect scientifically, of course it does not follow that it is also socially or morally unimportant or irrelevant, as it might have important consequences for society if it manifests itself after all. This already points to a possible tension between selecting risks for investigation from a scientific point of view and from a social or moral point of view.

The science of risk assessment also involves value judgments with respect to a number of methodological decisions that are to be made during risk assessment. Such methodological decisions influence the risk of error. A risk assessment might, due to error, wrongly estimate a certain risk or it might establish a risk where actually none exists. Heather Douglas (2009) argues that scientists in general have a responsibility to consider the consequences of error, just like anybody else. While this may seem common sense, it has important consequences once one takes into account the social ends for which risk assessments are used. Risk assessment is not primarily used to increase the stock of knowledge, but rather as an input for risk management. If a risk assessment wrongly declares something not to be a risk while it actually is a serious risk, or vice versa, this may lead to huge social costs, both in terms of fatalities and economic costs.

Various authors have therefore suggested that, unlike traditional science, risk assessment should primarily avoid what are called type 2 errors rather than type 1 errors (Cranor 1993; Shrader-Frechette 1991b; Hansson 2008; see also Hansson's "A Panorama of the Philosophy of Risk"). A type 1 error or false positive occurs if one establishes an effect (risk) where there is actually none; a type 2 error or false negative occurs if one does not establish an effect (risk) while there is actually an effect. Science traditionally focuses on avoiding type 1 errors to avoid assuming too easily that a certain proposition or hypothesis is true. This methodological choice seems perfectly sound as long as the goal of science is to add to the stock of knowledge, but in contexts in which science is used for practical purposes, as in the case of risk assessment, the choice may be problematic. From a practical or moral point of view it may be worse not to establish a risk while there is one than to wrongly assume a risk. As Cranor (1993) has pointed out the 95 % rule for accepting statistical evidence in science is also based on the assumption that type 1 errors are worse than type 2 errors. Rather than simply applying the 95 % rule, risk assessors might better try to reduce type 2 errors or balance type 1 against type 2 errors (Cranor 1993, pp. 32–29; Douglas 2009, pp. 104–106).

There are also other methodological decisions and assumptions that impact on the outcomes of risk assessment and the possibilities of error. One example is the extrapolation of empirically found dose–effect relations of potentially harmful substances to low doses. Often, no empirical data are available for low doses; therefore the found empirical data has to be extrapolated to the low dose region on the basis of certain assumptions. It might, for example, be assumed that the relation between dose and response is linear in the low dose region, but it is also sometimes supposed that substances have a no effect level, that is, that below a certain threshold dose there is no effect. Such methodological decisions can have a huge impact on what risks are considered acceptable. An example concerns the risks of dioxin. On basis of the same empirical data, but employing different assumptions about the relation between dose and response in the low dose region, Canadian and US authorities came to norms for acceptable levels of dioxin exposure to humans that are different by a factor of 1,000 (Covello and Merkhofer 1993, pp. 177–178).

While it is clear that in risk assessment, a number of value judgments and morally relevant methodological judgments need to be made, the implications for the responsibility of risk assessors, as scientists, are less obvious. One possibility would be to consider such choices to be entirely the responsibility of the risk assessors. This, however, does not seem like a very desirable option; although risk assessors without doubt bear some responsibility, it might be better to involve other groups as well, especially those responsible for risk management, in the value judgments to be made. The other extreme would be to restore the value-free science idea as much as possible. Risk assessors might, for example, pass on the scientific results including assumptions they made and related uncertainties. They might even present different results given different assumptions or different scenarios. While it might be a good idea to allow for different interpretations of scientific results, simply passing on all evidence to risk managers, who then can make up their mind does not seem desirable. Such evidence would probably be quite hard if not impossible to understand for risk managers. Scientists have a proper role to play in the interpretation of scientific data, albeit to avoid that data is deliberately wrongly interpreted for political reasons. Hence, rather than endorsing one of those two extremes, one should opt for a joint responsibility of risk assessors and risk managers for making the relevant value judgments while at the same recognizing their specific and different responsibilities. Among others, this would imply recognizing that risk assessment is a process that involves scientific analysis and deliberation (Stern and Feinberg 1996; Douglas 2009).

Individual Versus Collective Responsibility for Risks

When you get into your car in order to transport your children to school and yourself to an important work meeting, you expose a number of people to the risk of being injured or even killed in an accident. First, you expose yourself to that risk. Second, you expose your children to that risk. Third, you expose other drivers, passengers, pedestrians, and cyclists to that risk. Furthermore, someone made decisions that affected your driving: decisions about driving licenses, street lighting, traffic lights and signs, intersections, roundabouts, and so forth. Who is responsible for these different forms of risk exposure? There is an individual and a collective level at which to answer this question. The underlying philosophical question is that of individual and collective responsibility—to what extent and for which risks is an individual responsible and to what extent and for which risks is society collectively responsible? In the following, we will explain how these issues relate to each other. The analysis of road traffic serves as an example of how aspects of individual and collective responsibility reoccur in most areas of risk management and policy today.

The fact that you expose yourself to the risks associated with driving a car appears to be a primarily individual responsibility. As a driver with a license you are supposed to know what the relevant risks are. Unless you acted under compulsion or ignorance you are held responsible for your actions, in road traffic as elsewhere. As discussed in section Conceptions of Responsibility, the condition of voluntariness has been discussed by philosophers since Aristotle. When you voluntarily enter your car and know that you risk yours and others' health and life by driving your car, even if those risks are considered fairly small in probability terms, you are responsible in case something bad happens because you accepted the risks associated with driving. This assessment is, of course, complicated by the behavior of other road users. Perhaps someone else made a mistake or even did something intentionally wrong, thereby causing an accident. In that case, you are often considered responsible to some extent, because you were aware of the risks associated with driving and these risks include being exposed to other people's intentional and unintentional bad behavior. However, other road users may bear the greatest share of responsibility in case their part in the causal chain is greater and their wrongdoing is considered more serious. The point is that the individual perspective distributes responsibility between the individuals involved in the causal chain. The key elements are (1) individuals, (2) causation, and (3) wrongdoing. The one/s that caused the accident by doing something wrong is/are responsible for it. (In section Conceptions of Responsibility, we mentioned two further conditions for responsibility, i.e., freedom and knowledge. These are usually met in traffic accidents and therefore we do not mention them separately here, but they may be relevant in specific cases.) When attributing responsibility according to this approach the road transport system is taken for granted the way it is. However, as we noted, someone made decisions concerning the road transport system and the way you and your fellow road users are affected by those decisions.

The collective or systemic perspective, instead, focuses on the road transport system. Were the roads of a reasonable standard, was there enough street lighting, and was the speed limit reasonable in relation to the condition and circumstances of the road? The default is to look at what the individuals did and did not do and to take the road transport system as a given and this is often reflected in law. However, in some countries the policy is changing and moving toward a collective or systemic perspective. In 1997, the Swedish government made a decision which has influenced discussions and policies in other European countries. The so-called Vision Zero was adopted, according to which the ultimate goal of traffic safety policy is that no one is killed or seriously injured in road traffic (Nihlén Fahlquist 2006). This may be seen as obvious to many people, but can be contrasted to the cost-benefit approach according to which the benefit of a certain method should always be seen in relation to its cost. Instead of accepting a certain number of fatalities, it was now stated that it is not ethically justifiable to say that 300 or 200 are acceptable numbers of fatalities. In addition to this idea, a new view of responsibility was introduced. According to that approach, individuals are responsible for their road behavior, but the system designers are ultimately responsible for traffic safety. This policy decision reflected a change in perspective moving from individual responsibility to collective responsibility. Road traffic should no longer be seen purely as a matter of individual responsibility, but instead the designers of the system (road managers, maintainers, and the automotive industry) have a great role to play and a great share of responsibility for making the roads safer and saving lives in traffic. Instead of merely focusing on individuals, causation, and wrongdoing, the focus should be on (1) collective actors with the (2) resources and abilities to affect the situation in a positive direction. The example of road traffic illustrates how an activity often has a collective as well as an individual dimension. The adoption of Vision Zero shows that our views on who is responsible for a risky activity, with individual and collective dimensions, can be changed.

Furthermore, this example illuminates the difference between (1) backward-looking and (2) forward-looking responsibility. Sometimes when discussing responsibility, we may refer to the need for someone to give an account for what happened or we blame someone for what happened. In other situations we refer to the aim to appoint someone to solve a problem, the need for someone to act responsibly or to see to it that certain results are achieved. There are several distinctions to be made within these two broad categories, but it could be useful to make this broad distinction between backward-looking and forward-looking notions of responsibility (see also section Conceptions of Responsibility).

The issue of collective responsibility is a much discussed topic in contemporary philosophy. Some scholars argue that there is no such thing and that only individuals can rightly be considered responsible. This position was taken, for example, by Lewis (1948) some years after World War II and it is understandable that many people were skeptical to the idea of collective actors and collective guilt at that point in time. The world has changed a lot since then and 65 years after World War II the ideas of collective actors and holding collectives responsible are not as terrifying. On the contrary, against the background of multinational corporations, for example, banks and oil producers, behaving badly and causing harm to individuals it appears more and more crucial to find a way of holding such actors accountable for harm caused by them. Philosophers like Peter French have therefore defended the idea that collective agents, such as corporations or governments, can be morally responsible (e.g., French 1984). Some authors claim that collective responsibility is sometimes irreducible to individual responsibility, that is, a collective can be responsible without any of its members being responsible (French 1984; Gilbert 1989; Pettit 2007; Copp 2007). Others claim that collective responsibility is, in the end, analyzable only in terms of individual responsibility (Miller 2010). The collective responsibility of the government might, for example, be understood as the joint responsibility of the prime minster (as prime minister), other members of the government, members of the Parliament, and maybe civil servants. In section Further Research: Organizing Responsibility for Risks, we will explore possible tensions between individual and collective responsibility, and the so-called problem of many hands.

Scholars are likely to continue discussing whether the notion of collective responsibility makes philosophical sense and if so how it should be conceived. What cannot be denied is that in society we treat some risks as an individual responsibility and others as a collective responsibility. Whereas the risks associated with mountaineering are usually seen as individual responsibility, the risks stemming from nuclear power are seen as collective. However, it is arguably not always that simple to decide whether an individual or a collective is responsible for a certain risk. It is often the case that there is an individual as well as a collective dimension to risks. Climate risks arising from the emissions of carbon dioxide are good examples of this. Arguably, individuals have a responsibility to do what they can to contribute to the reduction of emissions, but governmental and international action is also crucial. Furthermore, it is also a matter of which notion of responsibility we apply to a specific context. While we sometimes blame individuals for having smoked for 40 years thereby causing their own lung cancer, we may make it a collective responsibility to give them proper care.

There are two general perspectives on the balance between individual and collective responsibility for health risks. First, the libertarian approach views lifestyle risks, for example, smoking, as an individual matter and relates causation to blame and responsibility for the cost of damage. A liberal welfare approach considers causation as one thing and paying for the consequences as another thing so that even if an individual is seen as having caused her own lung cancer, she should perhaps not have to pay for the health care she now needs. Furthermore, according to liberal welfare theories, individuals are always situated in a socio-economic context and, consequently, the fact that a particular individual smokes may not entirely be a matter of free choice. Instead, it may be partly due to the situation she is in, her socioeconomic context, education, and so forth, which entails a different perspective on causation, and hence also on the distribution of responsibility between the individual and the collective. The liberal welfare approach does not pay as much attention to free choice as the libertarian approach, or alternatively does not see choices as free in the same sense as libertarians do. This is because the two perspectives assume different conceptions of liberty. Libertarians focus on so-called negative freedom, that is, being free to do whatever one wants to do as long as one does not infringe on another person's rights. Liberal welfare proponents focus on positive liberty, that is, freedom to act in certain ways and having possibilities to act. The former requires legislation to protect indi-viduals' rights and the latter requires a more expansive institutional setting and taxation to create the circumstances and capabilities (see "The Capability Approach in Risk Analysis") needed for people to make use of those possibilities. Different conceptions of liberty entail different conceptions of responsibility. Those emphasizing negative liberty attribute a greater share of responsibility to individuals and those who prefer positive liberty make governments and societies collectively responsible to a greater extent. (For a classic explanation of the concepts of negative and positive liberty see Isaiah Berlin 1958).

The decision to view a certain risk as an individual or a collective matter entails different strategies for dealing with risk reduction and different strategies for deciding about the acceptability of a risk. If the risk is seen as an individual matter the strategy is likely to emphasize information campaigns at the most. If, for example, road safety is seen as an individual responsibility risk managers who want to reduce the number of fatalities and injuries will inform the public about risky behavior and how to avoid such behavior. "Don't drink and drive" campaigns is an example of that strategy. Some libertarians would probably argue that even this kind of campaign is unacceptable use of taxpayers' money and that an information campaign should only objectively inform about the risks of drunk driving and not give any advice because individuals should be considered competent enough to make their own decisions about driving. However, a "Don't drink and drive" campaign could also be seen as a way of making sure individuals do not harm each other, that is, do not infringe on other individuals' rights not to be harmed, and for this reason it would probably be acceptable to a moderate libertarian. Surely, libertarians would not agree to anything more intrusive than this, for example, surveillance cameras.

If, instead, road safety is seen as a collective responsibility, risk managers may try to find other ways of reducing the risks of driving. In the case of drunk driving, one such example could be alcohol interlocks, that is, a new technology which makes it impossible to drive under the influence of alcohol. This device measures the driver's blood alcohol concentration (BAC) before the car starts, for example, through an exhalation sample, and because it is connected to the car's ignition it will not start if the measured concentration is above the maximum set. Alcohol interlocks are currently used in some vehicles and some contexts in Sweden and elsewhere. It is possible that the device will be a natural part of all motor vehicles in the future and this would indeed be a way of making drunk driving a collective responsibility, although individuals would still be responsible for not misleading or otherwise circumventing the system.

The collective approach to responsibility for risks is sometimes criticized for being paternalistic. The argument is that people should be free to make their own decisions about which risks are worth taking. One way to assure freedom of choice is to apply the principle of informed consent to decisions about acceptable risk. Informed consent is a principle commonly used in medical experiments and the idea is that those who take part in the experiments are informed about the risks and then decide whether to consent through signing a document. Similarly, individuals are to decide what technological risks they want to take. To this end, they should be informed about the risks of different technologies, and they should be free to decide whether to take a certain risk or not. The approach of informed consent clearly fits in a libertarian approach to risk taking. However, when people make decisions about risks, their choices can be affected through the way information is presented. Thaler and Sunstein (2008) argue that a decision is always made in a context and that "choice architects" design this context. Since choices are always framed in one way or another, you might as well opt for "nudging" people in the "better", healthier for instance, direction. One example of this is a school cafeteria in which different food products are arranged in one way or another and without removing the less healthy options, a "choice architect" could nudge children in the direction of the healthier options. Even a very anti-paternalistic libertarian, they argue, could accept this since no options are removed and the food has to be arranged in one way or another (Thaler and Sunstein 2008).

There are, however, several problems with applying the principle of informed consent to risk taking. One problem is that it might be hard, if not impossible, to present risks in a neutral and objective way (see also section Risk Communication). Second, risks are sometimes uncertain. Imagine there is research on how radiation from mobile phones affects grown-ups in the time frame of 10 years after you start using the phone, but not how it affects children or how it affects grown-ups in the long-term perspective of say 20–30 years. When you use your mobile phone or you let your child use one and you have been informed about the known risks, have you consented to all risks of radiation stemming from mobile phones? Third, it might be doubted whether all risks are or can be taken voluntarily, take for example, the risks associated with driving in an area lacking public transport. Fourth, in many cases the decision whether to accept or take a certain risk is or cannot be an individual decision because it affects other people. Take for example, the decision whether a certain area of the Netherlands should be additionally protected against the sea given expectations of rising sea levels due to the greenhouse effect. Such measures are likely to be very costly and whereas some individuals will judge that an increased risk should be accepted rather than spending large of amounts of public funds on higher dikes, others are likely to make the opposite assessment.

Decisions about which risks of flooding should be accepted are by their very nature collective decisions. Since such collective decisions are usually based on majority decision making, individual informed consent is not guaranteed. An alternative would be to require consensus, to safeguard informed consent, but that would very likely result in a stalemate and in a perseverance of the status quo. That would in turn lead to the ethical issue of how the status quo is to be understood. For example, in the case of increased likeliness of flooding the question is whether the status quo should be understood in terms of the current risk of flooding, so that maintaining the status quo would mean heightening the dikes, or whether it should be understood in terms of the current height of the dikes and accepting a higher risk of flooding.

Many acts of seemingly individual risk taking have a collective element. Even committing suicide by driving or jumping in front of a train is not an individual act since other road users may come in the way and get hurt and there is probably psychological damage to the train driver and others who see it happen. By driving your car you inevitably risk the lives of others when you risk your own life. Your own risk taking is then intertwined with the risk exposure of others.

The upshot of the above discussion is not that all decisions about risk are or should be, at least partially, collective decisions, but rather that we should distinguish different kinds of risks, some more individual and others more collective. Consider, for example, the alleged health risks of radiation from mobile phones. The risk that is generated by using a mobile phone, and thereby exposing oneself to radiation, is an individual risk; the radiation only affects the user of the phone. Radiation from base stations, on the other hand, is a collective risk. This is why it has been suggested that the former is managed through informed consent whereas the latter should be subject to public participation and democratic decision making (IEGMP 2000).

However, even if we decide that the risks associated with using a mobile phone is sometimes an individual responsibility, it should be noted that a seemingly individual risk carries with it aspects of collective decision making and responsibility since the government and international agencies may have to set a minimal risk level (MRL) stating what is acceptable radiation and what is not. Many contemporary risks are complex and collective. As democratic societies we have to make choices about what risks to allow. There is a procedural dimension to this, but also a normative dimension. As noted by Ferretti (2010), scholars have been discussing how to make sure that the procedure by which decisions about risks are made become more democratic and fair, but that we also have to discuss the normative and substantive issues of what risks are acceptable and what the decisions are about.

Risk Communication

As we have seen the tasks of risk assessment, risk management and risk reduction involve different groups, such as engineers, scientists, the government, company managers, and the public, with different responsibilities. Since each group has its specific expertise and fulfilling one's specific responsibility often requires information from others, communication between the groups is of essential importance. Risk communication is therefore crucial for the entire system of dealing with risks in order to work.

In the literature, risk communication is often understood as communication between the government and the public (e.g., Covello et al. 1989) (see "Tools for Risk Communication" and "Emotion, Warnings, and the Ethics of Risk Communication"). Although as indicated it might be advisable to understand the notion of risk communication broader, we will here follow this convention and understand risk communication as the communication between the government (or a company) and the public. The goals of such risk communication depend to an important extent on whether one conceives of risk management as an individual or collective responsibility as discussed in the previous section. As we saw there, whether risk management is seen as an individual or collective responsibility partly depends on one's philosophical or political stance. However, it also depends on the kind of risks focused on. Moreover, as we argued, risks are often both an individual and collective responsibility. Therefore, the distinction between individual and collective responsibility does not exactly match comparable distinctions between consequentialist and deontological approaches or between liberal and paternalistic approaches.

If one conceives of risk management, and especially of decisions about acceptable risk, as the individual responsibility of the one taking or undergoing the risk, the responsibility of the government as risk communicator is to inform the public as completely and as accurately as possible. However, it seems that the government should refrain from attempts to convince the public of the seriousness or acceptability of risks. In this frame, the goal of risk communication is to enable informed consent and the responsibility of the risk communicator is basically to provide reliable and relevant information to enable informed consent.

However, if one conceives of decisions about acceptable risk and risk management as a collective responsibility, trying to convince the public of the acceptability or seriousness of certain risks or trying to get their cooperation for certain risk management measures is not necessarily or always morally problematic, especially if the risk communicator is open about his or her goals (cf. Morgan and Lave 1990; Johnson 1999; Thaler and Sunstein 2008). In a liberal society, it might in general be improper for the government to deliberately misinform the public or to enforce certain risk measures, but convincing the public is not necessarily morally problematic. Moreover, in some extreme situations even misinformation and enforcement might be considered acceptable. It is, for example, generally accepted that violence may sometimes be used by the police to reduce the risks of criminality and terrorism. With respect to risk communication, one might wonder whether it would be acceptable to be silent about the risk of burglary if people have to leave their homes as quickly as possible because of the safety risk as a result of a coming hurricane. Misinformation about risks may in some cases be deemed acceptable if the consequences, or risks, of proper information are larger than the risks communicated. In such cases, consequentialist considerations may be considered more relevant than deontological considerations. In general, if one conceives of risk management as collective rather than as a purely individual responsibility, the consequences of risk communication may be relevant to the responsibilities of the risk communicator and these responsibilities can thus extend beyond informing in the public as well as possible. However, it seems that if one accepts that some risk management decisions are a collective responsibility, one can still either take a more consequentialist or a more deontological view on risk communication.

It might seem that the question concerning what information to provide to the public only arises if risk management is (partly) seen as a collective responsibility. If risk management is an individual responsibility and the aim of risk communication is to enable informed consent, the risk communicator should simply pass on all information to the public. However, not all information is equally relevant for informed consent, and so a certain choice of filtering of information seems appropriate. In addition to the question of what information should be provided, ethical questions may arise in relation to the question of how the information is to be framed (Jungermann 1996).

Tversky and Kahneman (1981) have famously shown that the same statistical information framed differently leads to contradictory decisions about what risks are accepted, for example, depending on whether risk information is framed in terms of survival or death. There are many others factors that are relevant for how risks are presented. One issue is the risk measure used. It makes a difference whether you express the maximum dosage of dioxin per day in picograms, milligrams, or kilograms. The latter presentation—maybe unintentionally—gives the impression that the risk is far smaller than in the first case. Another important issue in risk communication is how uncertainty should be dealt with. Should the risk communicator just communicate the outcome of a risk assessment or also include uncertainty margins? Should the risk communicator explain how the risk assessment was carried out, so that people can check how reliable it is? Should the methodological assumptions and choices made in the risk assessment (section Conceptions of Responsibility) be explained to the public?

Further Research: Organizing Responsibility for Risks

A major philosophical challenge today is to conceptualize responsibility in relation to collective agency. While the increased control over the environment seems to increase the total amount of responsibility, this responsibility is also increasingly dispersed over many different individuals and organizations. The somewhat paradoxical result is that it sometimes appears to be increasingly difficult to hold someone responsible for certain collective effects like climate change. Partly this may result from the fact that today's society is so obsessed with holding people responsible (blameworthy) that many individuals and organizations try to avoid responsibility rather than to assume it. Ulrich Beck (1992) has described this phenomenon as "organized irresponsibility".

We have identified five important topics for further research that we will discuss in the following sections. In section The Problem of Many Hands (PMH), we discuss what has been called the problem of many hands (PMH).

In section Climate Change as an Example, we will discuss the risk of climate change as an example of the PMH. We are all contributing to climate change. However, if and how this observation of (marginal) causal responsibility has implications for moral responsibility is not at all clear and this issue needs considerable attention.

Section Responsibility as a Virtue will discuss the idea that rather than understanding responsibility in a formal way, we should appeal to individuals who should take up responsibility proactively. To that purpose, we suggest to turn to virtue ethics and care ethics. We explore the possibilities of an account of responsibility as the virtue of care, as a way to deal with the PMH.

Another example is the discussion in section Responsibility for Risks about the related responsibilities for risk assessment, risk management, risk reduction, and risk communication. We have seen that there are some problems with the traditional allocation of responsibilities in which, for example, scientists are only responsible for risk assessment and have no role to play in risk management. These examples illustrate the need to discuss the distribution of responsibility (section The Procedure of Responsibility Distribution) among the actors involved as well as the question of who is responsible for the entire system, which involves the notion of institutional design (section Institutional Design).

The Problem of Many Hands

Although Dennis Thompson (1980) already coined the term "problem of many hands" in 1980, relatively little research has been done in this area. For this reason, we will summarize briefly what already has been done, but large parts of the discussion relate to directions and suggestions for further research.

Thompson describes the PMH as "the difficulty even in principle to identify who is responsible for... outcomes" (Thompson 1980, p. 905). Many different individuals act in different ways and the joint effect of those actions is an undesired state-of-affairs X, but none of the individuals (1) directly caused X or (2) wanted or intended X. In such cases, it is either difficult to discern how each actor con-tributed to X or it is unclear what implications the joint causal responsibility should have for the moral responsibility of the individuals whose combined actions caused X. As we have seen, there is backward-looking and forward-looking responsibility. The PMH can be seen as a problem of forward-looking responsi-bility, but it has primarily been discussed as a problem of backward-looking responsibility. Typically, the PMH occurs when something has happened and although there may not have been any wrongdoing legally speaking, the public may have a feeling that something has been done for which someone is morally responsible. The question is just who should be considered responsible, since the traditional conditions of responsibility are extremely hard to apply.

Two features of contemporary society make the PMH salient today. First, human activities are to an increasing extent carried out by groups of people instead of by individuals. Second, we are increasingly able to control risks and hazards, which also seem to increase our responsibilities. We will briefly discuss these features in turn.

Traditionally, philosophers theorize about morality in relation to individuals and how they act. However, in contemporary societies, a substantial part of the daily lives of individuals are intertwined with collective entities like the state, multinational corporations, nongovernmental organizations, and voluntary asso-ciations. Collective agency has become frequent. We talk about nations going to war, companies drilling for oil, governments deciding to build a new hospital, a local Lions club organizing a book fair. As discussed in section Individual Versus Collective Responsibility for Risks, the concept of collective moral responsibility is a much debated topic in philosophy. A risky activity can be seen as an individual or a collective responsibility, but most risks have aspects of both.

Collectively caused harm complicates ethical analysis. This is so partly for epistemic reasons, that is, because we do not know how the actions of different individuals combine to cause bad things. Furthermore, did each and every indi-vidual in that particular collective know what they took part in? However, it is not merely for epistemic reasons that we have problems ascribing responsibility in such cases. Collective harm may also arise due to a tragedy of the commons (Hardin 1968). In a tragedy of the commons, the commons—a shared resource—are exhausted because for each individual it is rational to use the commons as much as possible without limitation. The aggregate result of these individual rational actions, the exhaustion of the common resource so that no individual can continue to use it, is undesirable and in a sense irrational. Many environmental problems can be understood as a tragedy of the commons. Johnson (2003) has argued that in a tragedy of the commons individuals are not morally required to restrict their use of the common resource as long as no collective agreement has been reached, hence no individual can properly be held responsible for the exhaustion of the commons (for a contra argument, see Braham and van Hees 2010).

Similarly, Pettit (2007) has argued that sometimes no individual can properly be held morally responsible for undesirable collective outcomes (for support of Pettit's argument see, e.g., Copp (2007), for criticism, see Braham and van Hees (2010), Hindriks (2009), and Miller (2007). The type of situations he refers to are known as voting paradoxes or discursive dilemmas. Pettit gives the following example (Table 5.1). Suppose that three employees (A, B, and C) of a company need to decide together whether a certain safety device should be installed and suppose that they agree that this should only be done if (1) there is a serious danger (p), (2) the device is effective with respect to the danger (q), and (3) the costs are bearable (r). If and only if all three conditions are met ($\hat{p}$ $\hat{q}$ r) the device is to be installed implying a pay sacrifice (s) for all three employees. Now suppose that the judgments of the three individuals on p, q, r, and s are as indicated in the table below. Also suppose that the collective decision is made by majority decision on the individual issues p, q, and r and then deducing s from ($\hat{p}$ $\hat{q}$ r). The result would be that the device is installed and that they all have to accept a pay sacrifice. But who is responsible for this outcome? According to Pettit neither A, B, or C can be properly be held responsible for the decision because each of them believed that the safety device was not worth the pay sacrifice and voted accordingly as can be seen from the table (based on the matrix in Pettit 2007, p. 197). Pettit believes that in cases like this the collective can be held responsible even if no individual can properly be held responsible. Like in the case of the tragedy of the commons, this suggests that the collective agency may make it impossible to hold individuals responsible for collective harmful effects.

In addition to the salience of collective agency today, in today's society negative consequences often result from risk rather than being certain beforehand. Whereas moral theories traditionally deal with situations in which the outcome is knowable and well determined, societies today spend a considerable amount of time and money managing risks, that is, situations in which there is a probability of harm. If it is difficult to decide whether killing is always wrong when done by and to individuals, it is even more difficult to decide whether it is acceptable to expose another human being to the risk of, say 1 in 18,000, of being killed in a road crash. Or, on the societal level, are the risks associated with nuclear power ethically acceptable? Would a difference in probabilities matter to the ethical acceptability and if so, where should the line be drawn between acceptable and unacceptable probability? It is difficult to know how to begin to answer these questions within the traditional ethical frameworks (Hansson 2009). The questions concerning the

Table 5.1 The discursive dilemma (based on Pettit 1997)

	Serious danger? (p)	Effective measure? (q)	Bearable costs? (r)	Pay sacrifice s(p̂ q̂ r)
A	No	Yes	Yes	No
B	Yes	No	Yes	No
C	Yes	Yes	No	No
Majority	Yes	Yes	Yes	(Yes) no

ethical acceptability of risks clearly have implications for responsibility. If A kills B, A is reasonably held responsible for it and the consequences of that vary according to norms and context. If A exposes B to the risk of 1 in 18,000 of being killed in a road crash, in what way is A responsible for that risk exposure? Interestingly, while it appears more intricate to decide how someone is responsible for exposing another person to the risk of dying than it is to decide whether someone is responsible for killing that person, the very concept of risk appears to imply some sense of responsibility. A risk is often seen as something we can or ought to be able to manage and control (cf. section Conceptual Relations Between Risk and Responsibility).

Thus, contemporary society is confronted by more collective agency and possibly more risks. These two features put the so-called problem of many hands (PMH) to the fore. A lot may be at stake: people's lives, the environment, and public health. Furthermore, in addition to cases where the probability is relatively well known, technological research and development entail substantial uncertainty about future hazards about which we do not have any knowledge today.

Climate Change as an Example

Climate change is an illustrative example of a substantial risk (or cluster of risks) for which it is extremely difficult to ascribe and distribute responsibility and which is caused by more or less all human beings, private companies, and governments. It is therefore a possible example of the PMH.

In debates about climate change, various notions of responsibility are at play (cf. section Conceptions of Responsibility) as is reflected in the different principles of responsibility that have been proposed. First, there is the polluter pays principle (PPP) stating that the polluting actor, that is, the one who caused the pollution, is the actor who ought to pay the cost (United Nations 1992; Caney 2005; Shue 1999). This principle applies a backward-looking notion of responsibility since it focuses on the causal link, but it also associates backward–and forward-looking responsibility in the claim that the one who caused the damage is also the one who should rectify the situation.

Second, there is a principle referred to as common, but differentiated responsibilities (CDR), which states that although all countries share responsibility for climate change, the developed nations have a greater share of responsibility to do something about it (forward-looking responsibility) because their past and current causal contribution is greater (backward-looking responsibility) (United Nations 1998). Thus, both the PPP and the CDR assume that the agent who caused climate change is also the one who is responsible to improve the situation. We often think about responsibility in these terms, but it is possible to conceive of responsibility for climate change differently. The ability to pay principle represents a different approach (Caney 2010). Originally, this principle is associated with a progressive tax system to justify why wealthy people should pay a greater share of their incomes in taxes than poor people in order to maintain a social welfare system. It is possible to design a principle of responsibility for climate change in a similar vein. A central principle in ethics is "'ought' implies 'can'" essentially meaning that it does not make sense to demand that people do X if they are unable to do X. It has also been argued that sometimes "'can' implies 'ought'" (Garvey 2008). This means that it may be reasonable to attribute a greater share of responsibility for climate change to developed nations not only because they contributed more to the causal chain, but because they simply have more resources to do something about it. This would of course not be reasonable for all risks, but considering the scope and potentially devastating consequences of this particular cluster of risks, it may be a reasonable principle in this case.

We have seen that there are different ways of approaching the distribution of responsibility for climate change between collective actors. In addition, there is the question about how to distribute responsibility between individuals versus collective agents, for example, governments and private companies. To what extent are individuals responsible? Furthermore, in what ways and for which parts are they responsible? Some philosophers argue that individuals are responsible, in the sense of accountability and blameworthiness (backward-looking responsibility) (e.g., Braham and van Hees 2010). Others argue that individuals are not responsible, but that governments are (e.g., Sinnott-Armstrong 2005), and still others argue that individuals are responsible in a forward-looking way, but that they are not to blame for how climate change and environmental problems came about (e.g., Nihlén Fahlquist 2009).

By talking about risks instead of direct harm, we have changed the perspective of time. A risk that something negative could happen is something for which someone can take responsibility and do something about. This is different from cases in which harm has already been done. When the risk has materialized, we want to find someone to blame or give an account of what happened. We need a backward-looking notion since harm will be done and we will want to blame someone to compensate victims. However we also need responsible engineering, research, and risk management, that is, people who act responsibly in order to minimize the risks to society, people, and the environment.

The typical PMH situation occurs when something undesirable has happened as a consequence of collective acting. The PMH can be described by the question: "Who did that?!" which is the epistemic problem of knowing who actually did something to cause the undesired event, but the PMH can also point to the normative problem that we cannot find anyone whom it would be fair to hold responsible for the undesired event. The responsibility notion assumed in this question appears to be individualistic and backward looking. Although this notion of responsibility is common and in some ways necessary, there are other notions which may complement it. After all, if we are interested in solving the PMH we probably have to look not only for ways to attribute blame when a risk has materialized, but also for ways in which risks can be reduced or managed in a responsible way to prevent them from materializing. Presumably, what we want is to prevent the PMH from occurring. In the following sections, we will look into three ways to, if not replace, supplement the "Who did that?!"-approach to responsibility: responsibility as a virtue (Responsibility as a Virtue), responsibility distributions (The Procedure of Responsibility Distribution), and institutional design (Institutional Design).

Responsibility as a Virtue

Responsibility is an unusually rich concept. Whereas many notions of responsibility focus on attributing blame for undesired events, there is also a notion that focuses on character traits and personality. To be responsible can be more than having caused X, being blameworthy for causing X, or even having particular obligations to do something about X. Responsibility can also be a virtue and a responsible person can be seen as a virtuous person, that is, having the character traits of a responsible person (see section Conceptions of Responsibility). We will now take a closer look at this virtue-ethical notion of responsibility.

By researching, developing, and using technology, opportunities are created. In this process, risks are created as well. In essence, technology creates opportunities and threats. It is, in this sense, a double-edged sword. For example, we want oil for energy, which means that we have to deal with risk of leakage as well as the actual leakage when it happens. Although we live with risks every day, it becomes clear to most people only when the risk actually materializes. Intuitively many people probably think that activities providing us with opportunities, but also risks, imply an increased sense of responsibility and that such activities should be carried out responsibly.

To associate the concept of responsibility with character traits and a "sense of responsibility" means having a closer look at virtue ethics (see "Risk and Virtue Ethics"). Virtue ethics is often mentioned as the third main branch of ethical theories (next to consequentialism and deontology). Virtue ethicists attempt to find answers to questions of what an agent should do by considering the agent's character and the morally relevant features of the situation (Van Hooft 2006, p. 21). Seeing responsibility as a virtue would entail a focus on how to develop and cultivate people's character with the aim to establish a willingness to actively take responsibility. A willingness to take responsibility involves emotions such as feeling personal involvement, commitment, and not leaving it to others, a feeling that it is up to me and a willingness to sacrifice something (Van Hooft 2006, p. 144, see "Moral Emotions as Guide to Acceptable Risk"). It is not the same as a willingness to accept blame for things an agent has done wrong (backward looking) although that may be one part (Van Hooft 2006, p. 141). The main focus is on forward-looking responsibility.

One important aspect of responsibility as a virtue is the recognition that being a responsible person is about carefully balancing different moral demands (Williams 2008, p. 459). Against the background of the different kinds of moral demands human beings face today, it may be difficult to point to one action which is the only right one. Instead a virtuous-responsible person uses her judgment and finds a way to respond and optimize, perhaps, the various demands. Against this background, it could be argued that in order to avoid PMH, we need virtuous-responsible people who use their judgment to form a balanced response to conflicting demands. This could be one way of counteracting the organized irresponsibility of contemporary society. The question is of course how such a society or organization can be achieved. Virtue ethicists discuss upbringing, education, and training as ways of making people more virtuous (Hursthouse 2000; Aristotle 2000).

As mentioned in section The Problem of Many Hands (PMH), there are two features of contemporary society which combine to put the PMH to the fore. First, collective agency is increasing. Second, the number of risks has increased, or at least our desire to control risks has grown stronger. A virtue approach to responsibility may counteract the problem of many hands in two ways, both related to the second feature. First, focusing on responsible people may prevent risks from materializing instead of distributing responsibility when it has already materialized (see "Risk and Virtue Ethics"). Responsible people are concerned about risks to human health and the environment because they care and they use their judgment to prevent such risks from materializing. Second, when risks do materialize responsible people will not do everything to avoid being blamed, but will take ownership of what happened and make sure the negative consequences are minimized. Whether the virtue notion of responsibility could also meet the challenge of increasing collective agency is less straightforward. It could be argued that the very tendency to have more collective agency counteracts responsibility as a virtue since people can hide behind collective agents. However, the collectivization could also be seen as making it ever more important to develop a sense of responsibility. Such a development would probably start with moral education and training of young children, something which virtue ethicists often suggest as a way to cultivate virtue. It would also require organizations that foster virtues and a sense of responsibility (see also section Institutional Design).

The Procedure of Responsibility Distribution

As mentioned earlier, the concept of responsibility is extraordinarily rich and refers to not one but many different notions. In addition to the difference between legal and moral responsibility, there are many different notions of moral responsibility. It is not surprising that people have different notions in mind and what may appear as conflicts about who is responsible for a certain state-of-affairs may sometimes primarily be misunderstandings due to lack of conceptual clarity. However, it is not merely conceptual lack of clarity which causes differences. People disagree about the normative issues involved, that is, how responsibility should be understood and distributed. This is true for people in general and surely holds for professionals as well. According to Doorn (2010, 2011), the prevalence of differences in views on responsibility may cause the PMH. One way of attempting to resolve these differences may be to focus on the procedural setting instead of the substantive conception of responsibility. In order to do this, it is important that we agree that there are disagreements. The solution is not to find and apply the right one, but rather to achieve respect for differences, consensus concerning the procedural setting (this may of course be hard to achieve), and possibly agreement on concrete cases of responsibility distributions.

In political philosophy, John Rawls famously showed that what is needed in pluralist societies is a consensus on the basic structure of society among different religious, moral, and other "comprehensive doctrines" (Rawls 1999 [1971], 1993). He argues that we cannot expect that all citizens in a pluralist society agree on politics, but there are some basic principles to which most reasonable people regardless of which doctrine they adhere to would agree, not the least because those very principles would grant them the right to hold those different doctrines. In order for people with different comprehensive doctrines to agree to a basic structure as being fair they justify it through working back and forth between different layers of considerations, that is, their (1) considered moral judgments about particular cases, (2) moral principles, and (3) descriptive and normative background theories. When coherence is achieved between these different layers, we have achieved a wide reflective equilibrium (WRE). In spite of differing judgments on particular cases, different moral principles, and background theories, people can justify the basic structure of society. When many people agree on the basic principles of fairness through different WRE we have an overlapping consensus. Therefore, even if the ways in which we justify it may differ substantially everyone agrees on something, that is, the basic structure of society.

Doorn applies Rawls' theory to the setting of R&D networks (see also Van de Poel and Zwart 2010). The aim is to develop a model which shows how engineers do not have to agree on a specific conception of responsibility as long as they agree on fair terms of cooperation. R&D networks are non-hierarchical and often lack a clear task distribution, which leaves the question of responsibility open. Doorn shows how a focus on the procedure for responsibility distribution instead of a substantive conception of responsibility makes it possible for engineers to agree on a specific distribution of responsibility. They can agree to it because the procedure was morally justified and fair, even if they disagree about a specific notion of responsibility. This way responsibility is distributed, that is, the PMH is avoided, but the professionals do not have to compromise their different views on responsibility. Without reaching a consensus on a responsibility notion or a responsibility distribution, consensus is reached on a procedure yielding legitimate responsibility distributions. In addition to Rawls' procedural theory there are others theories, for example, based on deliberative democracy that are set out by authors like Habermas, Cohen, and Elster which can be used in order to help focus on the procedure of distributing responsibility instead of the substantive notions (see, e.g., Habermas 1990; Bohman and Rehg 1997; Elster 1998).

Institutional Design

We have discussed responsibility as a virtue (Responsibility as a Virtue) and the procedure by which responsibility may be distributed (The Procedure of Responsibility Distribution) as two ways of counteracting the problem of many hands. We will now look at the importance of institutions. In particular, we will look at what has been called institutional design, the purposeful design of institutions (see, e.g., Weimer 1995). Since institutions generally already exist and cannot be designed from scratch, institutional design usually amounts to modulating or changing existing institutions. Institutional design may contribute to solving the PMH in two different ways: (1) it might create the appropriate institutional environment for people to exercise responsibility-as-virtue and (2) it might help to avoid unintended collective consequences of individual actions. We will discuss both possibilities briefly below.

Institutions may facilitate virtuous or vicious behavior. As argued by Hanna Arendt (1965), Eichmann was an ordinary person who, when he found himself in the context of Nazi-Germany, started to behave like an evil person. Institutions may socialize people into evil doing. Although most cases are not as dramatic and tragic as Eichmann's case, the institutions within and through which we act affect to what extent we act as responsible people. Larry May (1996) has developed a theory of responsibility that connects ideas about responsibility as a virtue to institutions. Institutions can facilitate and encourage or obstruct virtuous behavior. May discusses the ways in which our individual beliefs may change at the group level. The community has an important role in shaping the beliefs of individuals. Relationships between people require a certain collective consciousness with common beliefs. The important point about May's theory is that to foster a sense of responsibility-as-virtue among individuals in a group or organization requires an appropriate institutional environment. As we have argued before (Responsibility as a Virtue) fostering responsibility-as-virtue may contribute to solving the PMH. The additional point is that doing so not only requires attention for individuals and their education but also for their institutional environment.

The other way that institutional design can contribute to solving the PMH is by devising institutions that minimize unintended collective consequences of individual actions. As we have seen in The Problem of Many Hands (PMH), the PMH partly arises because the actions of individuals may in the aggregate result in consequences that were not intended by any of the individuals. The tragedy of the commons and the discursive dilemma were given as examples of such situations.

The phenomenon of unintended consequences is, however, much more general. The sociologist Raymond Boudon (1981) has distinguished between two types of systems of interaction. In what he calls functional systems, the behavior of individuals is constrained by roles. A role is defined by "the group of norms to which the holder of the role is supposed to subscribe" (Boudon 1981, p. 40). In interdependent systems, roles are absent but the actors are dependent on each other for the achievement of their goals. An ideal-typical example of an interdependent system is the free economic market. The tragedy of the commons in its classical form also supposes an interdependent system of interaction and the absence of roles, since the actors are not bound by any institutional norms.

According to Boudon, emergent, that is, collective, aggregate effects are much more common in interdependent systems than in functional systems. Reducing emergent effects can therefore often be achieved by organizing an interdependent system into a functional system. This can be done, for example, by the creation of special roles. The "invention" of the role of the government is one example. In cases of technological risks, one may also think of such roles as a safety officer or safety department within a company or directorate, or an inspectorate for safety within the government. Another approach might be to introduce more general norms as constraints on action. This is in fact often seen as the appropriate way to avoid a tragedy of the commons. In both cases new role responsibilities are created. Such role responsibilities are obviously organizational in origin, but they may entail genuine moral responsibilities under specific conditions, like, for example, that the role obligations are morally allowed and they contribute to morally relevant issues (see also Miller 2010).

Conclusion

There are many different conceptions of risk and psychologists and philosophers have pointed out the need to include more aspects than probabilities and consequences, or costs and benefits when making decisions about the moral acceptability of risks. It remains to be seen whether these additional considerations also need to be built into the very concept of risk.

In addition there are many different notions of responsibility, and the exact relation between risk and responsibility depends on how exactly both notions are understood. Still in general, both risk and responsibility often refer to undesirable consequences and both seem to presuppose the possibility of some degree of control and of making decisions that make a difference.

We started this chapter by mentioning the Deepwater Horizon oil spill in 2010. People were outraged when it occurred and, it seems, rightly so. It raised the issue of responsibility-as- blameworthiness because it appeared as though there had been wrongdoing involved. However, it also raised the issue of responsibility-as-virtue since a lot of people joined the work to relieve the negative consequences of the oil spill and to demand political action to counteract companies from exploiting nature and human beings.

As this example shows, there is not only backward-looking responsibility for risk but also forward-looking responsibility. We discussed in some details relevant forward-looking responsibilities that might be attributed to engineers, risk assessors, risk communicators, and risk managers. We also discussed that risks may be more or less seen as individual or collective responsibility.

We ended with discussing the problem of many hands (PMH), which seems a possible obstacle to taking responsibility for the risks in our modern society. We also suggested three possible ways for dealing with the PMH: responsibility-as-virtue, a procedural approach to responsibility, and institutional design. What is needed is probably a combination of these three approaches, but the discussion suggests that there is also hope in that people are able to unite and release a collective sense of responsibility.

References

Arendt H (1965) Eichmann in Jerusalem: a report on the banality of evil. Viking, New York

Aristotle (ed) (2000) Nicomachean ethics. Cambridge texts in the history of philosophy. Cambridge University Press, Cambridge

Asveld L, Roeser S (eds) (2009) The ethics of technological risk. Earthscan, London

Beck U (1992) Risk society; towards a new modernity. Sage, London

Berlin I (1958) Two concepts of liberty. Clarendon, Oxford

Bohman J, Rehg W (1997) Deliberative democracy: essays on reason and politics. MIT Press, Cambridge

Boudon R (1981) The logic of social action: an introduction to sociological analysis. Routledge & Kegan Paul, London

Bradbury JA (1989) The policy implications of differing concepts of risk. Sci Technol Hum Values 14(4):380–399. doi:10.1177/016224398901400404

Braham M, van Hees M (2010) An anatomy of moral responsibility, manuscript. Available at http://www.rug.nl/staff/martin.van.hees/MoralAnatomy.pdf

Caney S (2005) Cosmopolitan justice, responsibility, and climate change. Leiden J Int Law 18(4):747–775

Caney S (2010) Climate change and the duties of the advantaged. Crit Rev Int Soc Polit Philos 13(1):203–228

Copp D (2007) The collective moral autonomy thesis. J Soc Philos 38(3):369–388. doi:10.1111/j.1467-9833.2007.00386.x

Covello VT, Merkhofer MW (1993) Risk assessment methods: approaches for assessing health and environmental risks. Plenum, New York

Covello VT, McCallum DB, Pavlova MT, Task force on environmental cancer and heart and lung disease (U.S.) (1989) Effective risk communication: the role and responsibility of government and nongovernment organizations, vol 4, Contemporary issues in risk analysis. Plenum, New York

Cranor CF (1993) Regulating toxic substances. A philosophy of science and the law, Environmental ethics and science policy series. Oxford University Press, New York

Davis M (1998) Thinking like an engineer. Studies in the ethics of a profession. Oxford University Press, New York

Davis M (2012) "Ain't no one here but us social forces": constructing the professional responsibility of engineers. Sci Eng Ethics 18:13–34. doi:10.1007/s11948-010-9225-3

Doorn N (2010) A Rawlsian approach to distribute responsibilities in networks. Sci Eng Ethics 16(2):221–249

Doorn N (2011) Moral responsibility in R&D networks. A procedural approach to distributing responsibilities. Simon Stevin Series in the Philosophy of Technology, Delft

Douglas M (1985) Risk acceptability according to the social sciences, vol 11, Social research perspectives: occasional reports on current topics. Russell Sage, New York

Douglas HE (2009) Science, policy, and the value-free ideal. University of Pittsburgh Press, Pittsburgh

Douglas M, Wildavsky AB (1982) Risk and culture: an essay on the selection of technical and environmental dangers. University of California Press, Berkeley

Elster J (1998) Deliberative democracy. Cambridge studies in the theory of democracy. Cambridge University Press, Cambridge

Ferretti M (2010) Risk and distributive justice: the case of regulating new technologies. Sci Eng Ethics 16(3):501–515. doi:10.1007/s11948-009-9172-z

Fischer JM, Ravizza M (1998) Responsibility and control: a theory of moral responsibility. Cambridge studies in philosophy and law. Cambridge University Press, Cambridge

Frankfurt H (1969) Alternate possibilities and moral responsibility. J Philos 66:829–839

French PA (1984) Collective and corporate responsibility. Columbia University Press, New York

Garvey J (2008) The ethics of climate change. Right and wrong in a warming world. Continuum, London

Giddens A (1999) Risk and responsibility. Modern Law Rev 62(1):1–10

Gilbert M (1989) On social facts. International library of philosophy. Routledge, London

Habermas J (1990) Moral consciousness and communicative action. Studies in contemporary German social thought. MIT Press, Cambridge

Hansson SO (2008) Risk. The Stanford encyclopedia of philosophy (Winter 2008 Edition)

Hansson SO (2009) Risk and safety in technology. In: Meijers A (ed) Handbook of the philosophy of science. Philosophy of technology and engineering sciences, vol 9. Elsevier, Oxford, pp 1069–1102

Hardin G (1968) The tragedy of the commons. Science 162:1243–1248

Harris CE, Pritchard MS, Rabins MJ (2008) Engineering ethics. Concepts and cases, 4th edn. Wadsworth, Belmont

Hart HLA (1968) Punishment and responsibility: essays in the philosophy of law. Clarendon, Oxford

Hindriks F (2009) Corporate responsibility and judgement aggregation. Econ Philos 25:161–177

Hunter TA (1997) Designing to codes and standards. In: Dieter GE, Lampman S (eds) ASM handbook, vol 20, Materials selection and design. pp 66–71

Hursthouse R (2000) On virtue ethics. Oxford Univesity Press, Oxford

IEGMP (2000) Mobile phones and health (The Stewart report). Independent Expert Group on Mobile Phones

International Program on Chemical Safety (2004) IPCS risk assessment terminology, Harmonization project document, vol 1. World Health Organisation, Geneve

Johnson BB (1999) Ethical issues in risk communication: continuing the discussion. Risk Anal 19(3):335–348

Johnson BL (2003) Ethical obligations in a tragedy of the commons. Environ Values 12(3):271–287

Jungermann H (1996) Ethical dilemmas in risk communication. In: Messick M, Tenbrunsel AE (eds) Codes of conduct. Behavioral research into business ethics. Russell Sage, New York, pp 300–317

Lewis HD (1948) Collective responsibility. Philosophy 24(83):3–18

Martin MW, Schinzinger R (2005) Ethics in engineering, 4th edn. McGraw-Hill, Boston

May L (1996) The socially responsive self. The University of Chicago Press, Chicago

Miller S (2007) Against the collective moral autonomy thesis. J Soc Philos 38(3):389–409. doi:10.1111/j.1467-9833.2007.00387.x

Miller S (2010) The moral foundations of social institutions: a philosophical study. Cambridge University Press, New York

Morgan MG, Lave L (1990) Ethical considerations in risk communication practice and research. Risk Anal 10(3):355–358

National Research Council (1983) Risk assessment in the federal government: managing the process. National Academy Press, Washington

Nihlén Fahlquist J (2006) Responsibility ascriptions and vision zero. Accid Anal Prev 38: 1113–1118

Nihlén Fahlquist J (2009) Moral responsibility for environmental problems—individual or institutional? J Agric Environ Ethics 22(2):109–124. doi:10.1007/s10806-008-9134-5

NSPE (2007) NSPE code of ethics for engineers. National Society of Professional Engineers, USA, http://www.nspe.org/Ethics/CodeofEthics/index.html. Accessed 10 Sept 2010

Pettit P (2007) Responsibility incorporated. Ethics 117:171–201

Rawls J (1993) Political liberalism. Columbia University Press, New York

Rawls J (1999 [1971]) A theory of justice, Revised edn. The Belknap Press of Harvard University Press, Cambridge

Rayner S (1992) Cultural theory and risk analysis. In: Krimsky S, Golding D (eds) Social theories of risk. Praeger, Westport, pp 83–115

Renn O (1992) Concepts of risk: a classification. In: Krimsky S, Golding D (eds) Social theories of risk. Praeger, Westport, pp 53–79

Roeser S (2006) The role of emotions in the moral acceptability of risk. Saf Sci 44:689–700

Roeser S (2007) Ethical intuitions about risks. Saf Sci Monit 11(3):1–13

Rolf L, Sundqvist G Sociology of risk. pp 1001–1028 (in press)

Shrader-Frechette KS (1991a) Reductionist approaches to risk. In: Mayo DG, Hollander RD (eds) Acceptable evidence; science and values in risk management. Oxford University Press, New York, pp 218–248

Shrader-Frechette KS (1991b) Risk and rationality. Philosophical foundations for populist reform. University of California Press, Berkeley

Shue H (1999) Global environment and international inequality. Int Aff 75(3):531–545

Sinnott-Armstrong W (2005) It's not my fault: global warming and individual moral obligations. In: Sinnott-Armstrong W, Howarth RB (eds) Perspectives on climate change science, economics, politics, ethics. Elsevier/JAI, Amsterdam, pp 285–307

Slovic P (2000) The perception of risk. Earthscan, London

Stern PC, Feinberg HV (1996) Understanding risk. Informing decisions in a democratic society. National Academy Press, Washington

Thaler RH, Sunstein CR (2008) Nudge: improving decisions about health, wealth, and happiness. Yale University Press, New Haven

Thompson DF (1980) Moral responsibility and public officials: the problem of many hands. Am Polit Sci Rev 74(4):905–916

Thompson PB, Dean W (1996) Competing conceptions of risk. Risk Environ Health Saf 7: 361–384

Tversky A, Kahneman D (1981) The framing of decisions and the psychology of choice. Science 211(4481):453–458. doi:10.1126/science.7455683

United Nations (1992) Rio declaration on environment and development

United Nations (1998) Kyoto protocol to the United Nations framework convention on climate change

Van de Poel I (2011) The relation between forward-looking and backward-looking responsibility. In: Vincent N, Van de Poel I, Van den Hoven J (eds) Moral responsibility. Beyond free will and determinism. Springer, Dordrecht

Van de Poel I, Royakkers L (2011) Ethics, technology and engineering. Wiley-Blackwell

Van de Poel I, Van Gorp A (2006) The need for ethical reflection in engineering design the relevance of type of design and design hierarchy. Sci Technol Hum Values 31(3):333–360

Van de Poel I, Zwart SD (2010) Reflective equilibrium in R&D Networks. Sci Technol Hum Values 35(2):174–199

Van Hooft S (2006) Understanding virtue ethics. Acumen, Chesham

Wallace RJ (1994) Responsibility and the moral sentiments. Harvard University Press, Cambridge

Weimer DL (ed) (1995) Institutional design. Kluwer, Boston

Williams G (2008) Responsibility as a virtue. Ethical Theory Moral Pract 11(4):455–470

Wolff J (2006) Risk, fear, blame, shame and the regulation of public safety. Econ Philos 22(03):409–427. doi:10.1017/S0266267106001040

Index

S. Roeser et al. (eds.), *Essentials of Risk Theory*, SpringerBriefs in Philosophy,
DOI: 10.1007/978-94-007-5455-3, © The Author(s) 2013